THE ELOQUENT

JACQUELINE

KENNEDY ONASSIS

THE ELOQUENT

JACQUELINE

KENNEDY ONASSIS

A PORTRAIT IN HER OWN WORDS

with a Fifty-Minute DVD from A&E Biography

Edited by

BILL ADLER

William Morrow
An Imprint of HarperCollins Publishers

The frontispiece: Jackie at home in Georgetown. (© The Estate of Jacques Lowe—Woodfin Camp & Associates. Courtesy of the John Fitzgerald Kennedy Library, Boston, Massachusetts.)

Grateful acknowledgment is made to reprint the following excerpts: "The First Lady, She Tells Her Plans for the White House," © 1961 TIME Inc., reprinted with permission. From March 1979 *Ms. Magazine* interview, published courtesy of Gloria Steinem.

Grateful acknowledgement is made for permission to duplicate and enclose the award-winning documentary film *Jackie O: In a Class of Her Own,* © 1996 by Arts & Entertainment Television Networks. All rights reserved.

HarperCollins books may be purchased for educational, business, or sales promotional use. For information please write: Special Markets Department, HarperCollins Publishers Inc., 10 East 53rd Street, New York, NY 10022.

FIRST EDITION

Designed by Claire Vaccaro

Special thanks to Shawn Coyne, Web Stone, and Sara Wytrzes at Rugged Land L.L.C. and A&E Television Networks.

Printed on acid-free paper

Library of Congress Cataloging-in-Publication Data has been applied for

ISBN 0-06-073282-2

04 05 06 07 08 ❖/RRD 10 9 8 7 6 5 4 3 2 1

CONTENTS

INTRODUCTION

Jack and Jackie were America's royal couple, Camelot come to life. Jacqueline Kennedy brought beauty, grace, and intelligence to the White House. She had so affected us that even long after her husband's tragic assassination, she remained in the hearts of Americans. The courage she exhibited, the choices she made for her own well-being and that of her family, and the often glamorous life she lived inspired a nation. Jackie was one of a kind. Hers is a fascinating story, filled with the highest highs and the lowest lows. This is Jackie's life—in her own words.

—BILL ADLER

HIGHLIGHTS

I took the choicest bachelor in the Senate."

*

He's an idealist—without illusions."

*

All the talk over what I wear and how I fix my hair has me amused, but it also puzzles me. What does my hairdo have to do with my husband's ability to be president?"

*

[It's] as though I have just turned into a piece of public property. It's really frightening to lose your anonymity at thirty-one."

*

I'll be a wife and mother first, then first lady."

*

*I*f Jack proved to be the greatest president of the century and his children turned out badly, it would be a tragedy."

*

*M*y husband never made a sound. He had this sort of quizzical look on his face and his hand was up. I remember thinking he just looked as if he had a slight headache. And then he put his hand to his forehead and fell into my lap."

*

*J*ack was the love of my life. No one will ever know a big part of me died with him."

*

And it will never be that way again. There'll be great presidents again, but there'll never be another Camelot."

★

I think my biggest achievement is that, after going through a rather difficult time, I consider myself comparatively sane."

★

So many people hit the White House with their Dictaphone running. . . . I never even kept a journal. I thought, 'I want to live my life, not record it.'"

★

Like everybody else, I have to work my way up to an office with a window."

★

When you get written about a lot, you just think of it as a little cartoon that runs along at the bottom of your life—but one that doesn't have much to do with your life."

★

What has been sad for many women of my generation is that they weren't supposed to work if they had families. . . . What were they going to do when the children were grown—watch the raindrops coming down the windowpane?"

★

If you produce one book, you will have done something wonderful in your life."

THE EARLY YEARS

Childhood

When Jackie was just four years old, she, her new-born sister, Lee, and their nanny went for a stroll in Central Park. A short while into their walk, Jackie wandered off. Just as a police officer spotted her walking alone, she stepped up to him and said firmly, "My nurse and baby sister seem to be lost."

Even at an early age, Jackie loved to read and write verse. The following is an excerpt from a childhood poem written in anticipation of Christmas:

> *Christmas is coming*
> *Santa Claus is near*

Reindeer hooves will soon be drumming
On the roof tops loud and clear.

⭐

About her early reading choices, Jackie once said: "I lived in New York City until I was thirteen and spent the summers in the country. I hated dolls, loved horses and dogs, and had skinned knees and braces on my teeth for what must have seemed an interminable length of time to my family. I read a lot when I was little, much of which was too old for me. There were Chekhov and Shaw in the room where I had to take naps and I never slept but sat on the windowsill reading, then scrubbed the soles of my feet so the nurse would not see I had been out of bed. My heroes were Byron, Mowgli, Robin Hood, Little Lord Fauntleroy's grandfather, and Scarlett O'Hara."

⭐

Jackie also recalled learning French as a child while sitting around the dining room table: "When we were children our mother used to make us play a game. We sat at the table and every child had in front of them ten matches. Each time you said an English word, you'd throw a match away. [The winner held the last match.]"

★

On yet another defining moment in her childhood, Jackie has said, "I was a tomboy. I decided to learn to dance and I became feminine."

School Days

Young Jacqueline Bouvier attended Chapin (elementary) School on the Upper East Side in Manhattan, and as loved ones recall, she was often sent to see the headmistress. Her mother once asked, "What happens when you're sent to Miss Stringfellow?"

"Well, I go to the office and Miss Stringfellow says, 'Jacqueline, sit down. I've heard bad reports about you.' I sit down. Then Miss Stringfellow says a lot of things—but I don't listen."

★

She spent her high school days at Miss Porter's, where she once wrote to a Farmington friend: "I just know no one will ever marry me and I'll end up as a house mother at Farmington."

★

When she graduated from Miss Porter's in 1947 at age eighteen, she wrote in the class yearbook under "Ambition in Life": "Not to be a housewife."

★

Jacqueline referred fondly to Miss Helen Shearman, who taught Latin at the Holton-Arms School, where Jacqueline studied for two years, and who also had a reputation for being somewhat demanding: "But she was right. We were all lazy teenagers. Everything she taught me stuck, and though I hated to admit it, I adored Latin."

★

All my great interests—in literature and art, Shakespeare and poetry—were formed because I was fortunate enough to find superb teachers in these fields."

*

Reminiscing about her transition into adulthood, she said, "It happened gradually over the three years I spent at boarding school trying to imitate girls who had callers every Saturday. I passed the finish line when I learned to smoke, in the balcony of the Normandie theater in New York from a girl who pressed a Longfellow upon me then led me from the theater when the usher told her that other people could not hear the film with so much coughing going on. Growing up was not so hard."

College Days

Jacqueline Bouvier attended Vassar College for two years, then spent her junior year at the Sorbonne in Paris and her senior year at George Washington University in Washington, D.C.

*

During a summer session at the University of Grenoble (where she was enrolled in a six-week intensive language-arts course prior to her year at the Sorbonne), she lived with a French family and wrote home: "They just grow on you so—they get nicer every day and open up to us and treat us like members of the family. We all laugh hysterically through meals and the mother is so good-natured. They are of the old aristocracy and hard up now and have to take in students."

*

She also recalled the kindness of the French students at the University of Grenoble:

"They helped us with our compositions. I wrote mine in halting French and my friend did it all over. It really was very hard and they all took it so seriously and searched for words to use—it was really so nice of them to take all that trouble with some dumb foreigner who couldn't do her homework."

✳

At age twenty-one, Jackie entered a contest, sponsored by *Vogue,* for a chance to win a trip to Paris. The following is an excerpt from the essay she submitted describing how her prior travels through Europe had influenced her:

"Being away from home gave me a chance to look at myself with a jaundiced eye. I learned not to be ashamed of a real hunger for knowledge, something I had always tried to hide, and I came home glad to start in here again but with a love for Europe that I am afraid will never leave me."

✳

While studying at the Sorbonne, she and Claude de Renty, the daughter of her landlady, took a trip together in 1950: "I had the most terrific vacation in Austria and Germany. We really saw what it was like with the Russians with Tommy guns in Vienna. We saw Vienna and Salzburg and Berchtesgaden where Hitler lived: Munich and the Dachau concentration camp. . . ."

Jackie's passion for horses began in her youth. She's pictured here riding in Middleburg, Virginia. (Photo by Robert Knudson. Courtesy of the John Fitzgerald Kennedy Library, Boston, Massachusetts.)

About that same trip, she wrote: "It's so much more fun traveling second and third class and sitting up all night in trains, as you really get to know people and hear their stories. When I traveled before it was all too luxurious and we didn't see anything."

*

They made several side trips in southern France:

"I just can't tell you what it is like to come down from the mountains of Grenoble to this flat, blazing plain where seven-eighths of all you see is hot blue sky—and there are rows of poplars at the edge of every field to protect the crops from the mistral and spiky short palm trees with blazing red flowers growing at their feet. The people here speak with the lovely twang of the 'accent du Midi.' They are always happy as they live in the sun and love to laugh. It was heartbreaking to only get such a short glimpse of it all—I want to go back and soak it all up. The part I want to see is la Camargue—a land in the Rhone delta which is flooded by the sea every year and they have a ceremony where they all wade in on horses and bless it—La Bénédiction de la Mer—gypsies live there and bands of little Arab horses and they raise wild bulls."

Jackie's Mother

Jackie's mother, Janet Lee, married Jackie's father, John B. "Black Jack" Bouvier, on July 7, 1928. Jacqueline Bouvier was born a year later, on July 28, 1929, at Southampton Hospital on Long Island. Her sister, Caroline Lee (called Lee), was born three and a half years later on March 3, 1933. The marriage foundered, and the parents separated in 1936, attempted a short-lived reconciliation in 1937, and finally divorced in 1940. In 1942 Janet married Hugh D. Auchincloss. The union produced two children, Janet, born in 1945, and Jamie, born in 1947. Auchincloss also had three children from previous marriages, Hugh III (Yusha), Nina, and Thomas.

★

During Jackie's junior year at the Sorbonne, she told her stepbrother Hugh Auchincloss: "I have to write Mummy a ream each week or she gets hysterical and thinks I'm dead or married to an Italian."

When her mother and stepfather had been married for a decade, Jacqueline, then twenty-three, wrote a series of poems, each spotlighting an event made possible only by their marriage. Her introduction read: "It seems so hard to believe that you've been married ten years. I think they must have been the very best decade of your lives. At the start, in 1942, we all had other lives and we were seven people thrown together, so many little separate units that could have stayed that way. Now we are nine—and what you've given us and what we've shared has bound us all to each other for the rest of our lives."

Jackie's Father

John B. "Black Jack" Bouvier was a stockbroker whose finances were as erratic as the stock market. Jacqueline adored him, and they maintained a close relationship, even after her parents separated and then divorced.

Bouvier often visited Jacqueline at Miss Porter's School: "All my Farmington friends loved Daddy. He'd take batches of us out to luncheon at the Elm Tree Inn. Everybody ordered steaks and two desserts. We must have eaten him broke."

The following is a poem Jackie had written for her father:

> *I love walking on the angry shore*
> *To watch the angry sea*
> *Where summer people were before,*
> *And now there's only me.*

★

When Jackie's father died in 1957, she planned his funeral at St. Patrick's Cathedral. She chose garlands of daisies in white wicker baskets, saying, "I want everything to look like a summer garden."

Jackie's Sister, Lee

Jackie said of her sister: "Lee was always the pretty one. I guess I was supposed to be the smart one."

★

Jacqueline also once complained to a friend: "Lee is so dippy about Jack it's sickening."

★

But she was also quoted as having said about Lee, "Nothing could ever come between us."

Early Career

I always wanted to be some kind of writer. . . . Like a lot of people, I dreamed of writing the Great American Novel."

*I*n 1952, while working as "The Inquiring Photographer" at the *Washington Times-Herald,* Jacqueline Bouvier, then twenty-two, wrote to newspaperwoman Bess Furman: "I'm so in love with all that world now—I think I look up to newspaper people the way you join movie star fan clubs when you're ten years old."

*S*he liked to interview children because "they make the best stories." She once interviewed Tricia Nixon, then six. It was right after the 1952 elections, and Nixon had just been elected vice president. Jackie's question was, "What do you think of Senator Nixon now?"

(Tricia answered: "He's always away. If he's famous, why can't he stay at home?")

*

On why she chose to study journalism: "I wanted to know people better. I thought studying journalism would be a great chance."

*

[I am concerned by] the myth that I am just a sheltered socialite. I proved that I could support myself by holding down a newspaper job for a year and a half and by winning the *Vogue* Prix de Paris."

*

Being a journalist seemed the ideal way of both having a job and experiencing the world, especially for anyone with a sense of adventure. I wouldn't choose it as a profession now—journalism has variety but doesn't allow you to enter different worlds in depth . . . though I understand why so many young people are attracted to it. Being a reporter seems a ticket out to the world."

In her coverage of the 1953 Eisenhower inauguration, Jackie's reporting paid particular attention to the first ladies: "Mamie's lively laughter could be heard far back in the crowd . . . while Mrs. Truman sat stolidly with her gaze glued on the blimp overhead through most of the ceremony. . . . Ike planted a kiss on Mamie's cheek right after taking the oath."

In her job as the Inquiring Photographer for the *Times-Herald,* her questions frequently showed a whimsical turn of mind:

"Do the rich enjoy life more than the poor?"

"Chaucer said that what women most desire is power over men. What do you think women desire most?"

"Do you think a wife should let her husband think he's smarter than she is?"

"If you were going to be executed tomorrow morning, what would you order for your last meal on earth?"

"Would you like to crash high society?"

"How do you feel when you get a wolf whistle?"

"Are men braver than women in the dentist's chair?"

★

*I*n retrospect, some of her queries seem almost prophetic:

"Which first lady would you most like to have been?"

"Would you like your son to grow up to be president?"

"Should a candidate's wife campaign with her husband?"

"If you had a date with Marilyn Monroe, what would you talk about?"

"What prominent person's death affected you most?"

✳

When she interviewed Senator Kennedy (whom she was dating at the time) for her column, she asked, "Can you give any reason why a contented bachelor would want to get married?"

Early Social Life

Before she met Jack Kennedy, Jacqueline Bouvier attended her fair share of parties and dances:

"They were okay. But Newport—when I was about nineteen, I knew I didn't want the rest of my life to be there. I didn't want to marry any of the young men I grew up with—not because of them but because of their life. I didn't know what I wanted. I was still floundering."

*

Exercising her trademark charm, she told dinner companions that the most important thing to her about a man "is that he must weigh more and have bigger feet than I do."

*

On her ideal man: "I look at a male model and am bored in three minutes. I like men with funny noses, ears that protrude, irregular teeth, short men, skinny men, fat men. Above all, he must have a keen mind."

*

During her Inquiring Photographer days, John Husted (her fiancé before she met Jack Kennedy) took her to meet his family. When Husted's mother offered her a baby picture of him, Jackie declined, saying: "If I want a picture of John, I'll take my own."

*

Separated by geography (Jacqueline worked in Washington, Husted in New York), they corresponded regularly and saw each other on weekends. Jackie had met Congressman Kennedy by this time and subsequently wrote to Husted: "Don't pay any attention to any of the drivel you hear about me and Jack Kennedy. It doesn't mean a thing."

Shortly after this, they broke off their engagement.

LIFE WITH JOHN F. KENNEDY

Courtship

Jack and Jackie first met at the home of Mr. and Mrs. Charles Bartlett in 1951. Congressman Kennedy was planning his campaign for the Senate the following year, and Jacqueline Bouvier had just graduated from George Washington University. The Bartletts brought them together again the following year when Kennedy was actively campaigning for the Senate and Jackie was working as the Inquiring Photographer. She would later explain: "I met him at the home of friends of ours who had been shamelessly matchmaking for a year, and usually that doesn't work out, but this time it did, so I am very grateful to them."

*

Right after Kennedy's election to the Senate in 1952, her cousin John H. Davis asked her if there was truth to the rumors that she and Jack were serious about each other. She laughed and began describing the senator-elect:

"You know, he goes to a hairdresser almost every day to have his hair done.

"And you know, if, when we go out to some party or reception or something, nobody recognizes him, or no photographer takes his picture, he sulks afterwards for hours.

"Really. He's so vain I can't believe it.

"Oh, sure, he's ambitious all right, he even told me he intends to be president someday."

And she tossed her head back and laughed again.

*

Further on the subject of her "spasmodic" courtship with Jack Kennedy: "He'd call me from some oyster bar up on the Cape with a great clink-

ing of coins, to ask me out to the movies the following Wednesday."

Ultimately she would say: "Jack was something
special and I know he saw something special in me
too. I remember my mother used to bring around all
these beaus for me but he was different."

*

As for Kennedy's relationship with Jackie's mom,
Jackie said, "I'm the luckiest girl in the world.
Mummy is terrified of Jack because she can't push
him around at all."

*

On the subject of the smooth relationship that had
developed between Jackie's dad and Kennedy, she
once remarked, "They talked about sports, politics,
and women—what all red-blooded men like to talk
about."

*

Jackie also confided to a friend that she was attracted to Kennedy because he was "dangerous, just like Black Jack."

*

During their courtship, she said to a friend, "I don't know if I'll live long enough to marry him."

*

Her concerns about their early relationship were also reflected in her writing. Here she speaks of herself in the third person:

> [She] knew instantly that he would have a profound, perhaps disturbing influence on her life. In a flash of inner perception, she realized that here was a man who did not want to marry. She was frightened. Jacqueline, in the revealing moment, envisaged heartbreak, but just as swiftly determined that heartbreak would be worth the pain.

Jackie and Jack set sail during the weekend of their engagement announcement. (Courtesy of The Standard-Times.*)*

In 1953 Jackie met Zsa Zsa Gabor by chance on a plane flight back from the coronation of Queen Elizabeth. Zsa Zsa recalled Jackie saying to her as they arrived at the airport:

"There's a young man who's going to propose to me."

Unknown to Jackie, Kennedy had once been involved with Zsa Zsa. The actress got off the plane first and greeted him in the waiting room. He introduced the two women.

"Miss Bouvier and I spent hours together on the plane," Zsa Zsa told Kennedy. "She's a lovely girl. Don't dare corrupt her, Jack."

"But he already has," whispered Jackie.

Senator Kennedy disproved Jackie's theory that he was not the marrying kind when he proposed to her later that year. However, Jackie chose not to go public with the announcement of their engagement right away. She told a close relative: "Aunt Maudie, I just want you to know that I'm engaged to Jack Kennedy.

But you can't tell anyone for a while because it wouldn't be fair to the *Saturday Evening Post*."

When asked what the *Saturday Evening Post* had to do with it, she answered: "The *Post* is coming out tomorrow with an article on Jack. And the title is on the cover. It's 'Jack Kennedy—the Senate's Gay Young Bachelor.'"

Asked by a reporter at the time of their engagement if she felt she had much in common with Kennedy, she replied candidly: "Since Jack is such a violently independent person, and I too am so independent, this relationship will take a lot of working out."

★

On the subject of Jack meeting her extended family she said: "Wait till I introduce Jack Kennedy to Aunt Edie [Edith Bouvier Beale, an eccentric who owned forty cats]. You know, I doubt if he'd survive it. The Kennedys are terribly bourgeois."

✴

Shortly after Jack proposed, Jackie's mother invited Rose Kennedy to Newport for lunch to discuss wedding plans. Jack and Jackie were also there. Before lunch they drove to Bailey's Beach Club. The bride-to-be reported:

"The two mothers were in front of the car, and we were sitting in the backseat, sort of like two bad children. Anyway, Jack and I went swimming. I came out of the water earlier; it was time to go for lunch, but Jack dawdled. And I remember Rose stood on the walk and called to her son in the water, 'Jack! . . . Ja-a-ack!' and it was just like the little ones who won't come out and pretend not to hear their mothers calling—'Jaack!' but he wouldn't come out of the water. I can't remember whether she started down or I went down to get him, but he started coming up, saying, 'Yes, Mother.'"

✴

The bride-to-be beamed: "I took the choicest bachelor in the Senate."

★

Before their wedding she proclaimed: "What I want more than anything else in the world is to be married to him."

Early Years of Marriage

I brought a certain amount of order to his life. We had good food in our house—not merely the bare staples that he used to have. He no longer went out in the morning with one brown shoe and one black shoe on. His clothes got pressed and he got to the airport without a mad rush because I packed for him. I can be helpful packing suitcases, laying out clothes, rescuing lost coats and luggage. It's those little things that make you tired.

"The thing that gives me the greatest satisfaction is making the house run absolutely smoothly so that Jack can come home early or late and bring as many unexpected guests as he likes. Frankly, this takes quite a bit of planning.

"During our first year of marriage we were like gypsies living in and out of a suitcase. It was turbulent.

Jack made speeches all over the country and was never home more than two nights at a time. To make matters even more restless, we had rented a house in Georgetown for six months, and when the lease ran out, we moved to a hotel.

"We spent the summer, off and on, at Jack's father's house in Hyannis Port. Ours was the little room on the first floor that Jack used to have by himself. It didn't take me long to realize it was only big enough for one."

∗

That first year I longed for a home of our own. I hoped it would give our lives some roots, some stability. My ideal at that time was a normal life with my husband coming home from work every day at five. I wanted him to spend weekends with me and the children I hoped we would have."

∗

When an interviewer asked Mrs. Kennedy in 1954 her theories for a successful marriage, she grimly responded, "I can't say I have any yet."

36

⭐

She quickly adjusted to the pace of life with Jack: "It was hectic but I loved it. You don't really long for a home of your own unless you have children."

⭐

On the subject of housekeeping, she said: "House-keeping is a joy to me. When it all runs smoothly, when the food is good and the flowers look fresh, I have much satisfaction. I like cooking but I'm not very good at it. I care terribly about food, but I'm not much of a cook."

⭐

While Jackie did much to support Kennedy's political career, her refusal to cultivate important people was reportedly a source of annoyance to him. At a party, he was overheard saying, "The trouble with you, Jackie, is that you don't care enough about what people think of you."

She snapped back, "The trouble with you, Jack,

is that you care too much about what people think of you."

*

When Jack underwent spinal surgery in October 1954, Jackie became his political aide, handling much of his personal correspondence. She wrote warm notes to dignitaries such as Lyndon Johnson:

"I just wanted to tell you how terribly much your kind letter meant to Jack. . . . I've just realized here I have been scribbling away about my husband's illness and never told you how wonderfully thrilled we are for you being [elected] Majority Leader— You must be so happy and proud—and I know that you will absolutely make history in it."

*

Referring to Kennedy's recuperation from his back operation in 1954-55: "I think convalescence is harder to bear than great pain."

*

While Kennedy was in New York Hospital after his back operation, Jackie took part in a prank in which actress Grace Kelly posed as a night nurse: "When Jack opened his eyes, he thought he was dreaming. He was hardly strong enough to shake hands with her. He couldn't even talk."

During his convalescence, Kennedy worked on his Pulitzer Prize–winning book *Profiles in Courage.* Jackie said of the experience: "This project saved his life. It helped him channel all his energies while distracting him from pain."

Commenting on the book: "A sense of history and ability to learn from the past is of prime importance to any man in a position of leadership today."

By 1956, after nearly three years of marriage, she told a reporter: "I wouldn't say that being married to a very busy politician is the easiest life to adjust to. But you think about it and figure out the best way to do things—to keep the house running smoothly, to spend as much time as you can with your husband and your children —and eventually you find yourself well adjusted. . . . The most important thing for a successful marriage is for a husband to do what he likes best and does well. The wife's satisfaction will follow."

✶

In 1977 Virginia Bohlin of the *Boston Globe* recalled and quoted from an interview given by Jackie at Hammersmith Farm on June 24, 1953, shortly before the Kennedys' wedding:

"I'm dying to get a place of our own, so I can fix it up myself and get our wedding gifts out of storage. What I hope we'll find is some little Georgetown house. I'd love to have a little cozy house you can really run yourself. Then to furnish it, I'd like some lovely comfortable things mixed in with some nice old pieces of furniture."

Four years later, in 1957, they bought a house in

Georgetown and moved in when Caroline was three weeks old: "I love our home in Washington. There has always been a child in it. . . . My sweet little house leans slightly to one side, and the stairs creak."

★

She once compared herself and Jack Kennedy to "icebergs," whose real selves stayed hidden. This "was a bond between us."

★

She didn't see her husband as a typical politician: "He's an idealist—without illusions."

★

He is a rock, and I lean on him for everything. He is so kind. Ask anyone who works for him! And he's never irritable or sulky. He would do anything I wanted or give me anything I wanted."

★

As for Jack's reputation as a glamour boy: "It's non-sense. Jack has almost no time anymore for sailboats and silly things. He has this curious, inquiring mind that is always at work. If I were drawing him, I would draw a tiny body and an enormous head."

The Kennedys

On her introduction to the Kennedy family, Jackie once said, "Just watching them wore me out."

✳

Commenting on the roughness of their sports games, she told her sister: "They'll kill me before I ever get to marry him. I swear they will."

✳

On her first meeting with the Kennedy clan en masse in Hyannis Port, Cape Cod: "How can I explain these people? They were like carbonated water, and other families might be flat. They'd be talking about so many things with so much enthusiasm. Or they'd be playing games. At dinner or in the living room, anywhere, everybody would be talking about something. They had so much interest in life—it was so stimulating. And so gay and so open and accepting.

"They even compete with each other in conversation to see who can say the most and talk the loudest."

Responding to the need of Jack's family to be perfect in everything they did, she commented early in their courtship about the game of tennis: "It was enough for me to enjoy the sport. It wasn't necessary to be the best."

Jackie never did get the hang of the Kennedys' touch football games. For her husband's sake, she tried for a while, but she was doomed from the moment she asked, "If I get the ball, which way do I run?"

★

The day you become engaged to one of them is the day they start saying how "fantastic" you are, and the same loyalty they show to each other they show to their in-laws. They are all so proud when one of them does well.

"They seem proud if I read more books, and of the things I do differently. The very things you would think would alienate them bring you closer to them."

★

She later remembered the Kennedy family (from *Times to Remember* by Rose Fitzgerald Kennedy):

"Something so incredible about them is their gal-lantry. You can be sitting down to dinner with them and so many sad things have happened to each, and—God!—maybe even some sad thing has happened that day, and you can see that each one is aware of the other's suffering. And so they can sit down at the table in a rather sad frame of mind. Then each one will begin to start to make

this conscious effort to be gay or funny or to lift each other's spirits; and you find that it's infectious, that everybody's doing it. They all bounce off each other. . . .

"They have been such a great help to me. My natural tendency is to be rather introverted and solitary and to retreat into myself and brood too much. But they bring out the best. No one sits and wallows in self-pity. It's just so gallant that it really makes you proud. And you think, look at these people and the effort they are making, and you think that's a lesson you want to take with you."

JOSEPH P. KENNEDY SR.

Jacqueline Kennedy became good friends with Joe Kennedy while in Palm Beach during her husband's convalescence from back surgery. After her husband and her father, she said of Joe, "I love him more than anybody—more than anybody in the world."

✶

I used to tell [Joe Sr.] he had no nuances, that everything with him was either black or white, while life was so much more complicated than that. But he never got angry with me for talking straight to him; on the contrary, he seemed to enjoy it."

While the others played football, Joe and Jackie would sit on the porch "talking about everything from classical music to the movies."

She found the elder Kennedy's old-fashioned slang particularly amusing: "[He] ought to write a series of grandfather stories for children, like 'The Duck with Moxie' and 'The Donkey Who Couldn't Fight His Way Out of a Telephone Booth.'"

She also defended Joseph Kennedy against charges that he ran his children's careers: "You'd think he was a mastermind playing chess, when actually he's a nice old gentleman we see at Thanksgiving and Christmas."

*T*ime magazine reported in 1956 that Joe Kennedy had offered Jackie a million dollars to stay married to Jack. When she saw the story, she phoned Joe and said, "Why only one million dollars, why not ten million?"

ROSE KENNEDY

*O*n her first meeting with Rose Kennedy: "The first time I met her was about a year, a little more than a year, before I married Jack, when I came that summer for a weekend. I remember she was terribly sweet to me. For instance, I had a sort of special dress to wear to dinner—I was more dressed up than his sisters were, and so Jack teased me about it, in an affectionate way, but he said something like, 'Where do you think you're going?' She said, 'Oh, don't be mean to her, dear. She looks lovely.'"

★

*R*emembering Rose Kennedy (from *Times to Remember* by Rose Fitzgerald Kennedy):

> When I married Ari [Onassis], she of all people was the one who encouraged me. Who said, "He's

*a good man." And, "Don't worry, dear." She's been
extraordinarily generous. Here I was, I was married to her son and I have his children, but she
was the one who was saying, if this is what you
think is best, go ahead. It wouldn't surprise anyone who has really known her; but anyway, how
extraordinarily generous that woman is in spirit.
I always called her "Belle Mère"—and I still call
her that . . .*

*If I ever feel sorry for myself, which is a
most fatal thing, I think of her. I've seen her cry
just twice, a little bit. Once was at Hyannis Port,
when I came into her room, her husband was ill,
and Jack was gone, and Bobby had been killed . . .
and the other time was on the ship after her husband died, and we were standing on deck at the
rail together, and we were talking about something . . . just something that reminded her. And
her voice began to sort of break and she had to
stop. Then she took my hand and squeezed it and
said, "Nobody's ever going to have to feel sorry for
me. Nobody's ever going to feel sorry for me," and
she put her chin up. And I thought, God, what a
thoroughbred.*

⋆

An old friend recalled a dinner in Paris, while Jackie was married to Onassis, that included Rose Kennedy. Jackie insisted that Ari take them to a nightclub: "You know, Rose hasn't been to a night-club since Joe took her to the Lido in 1936."

ROBERT F. KENNEDY

After the birth of her stillborn daughter in 1956, Bobby Kennedy was there for Jackie; her husband had been aboard a boat off the coast of Italy at the time. Many years later, she learned that Bobby had also arranged for the burial of the child: "You knew that, if you were in trouble, he'd always be there."

⋆

Hyannis, 1960: "Bobby is immensely ambitious and will never feel that he has succeeded in life until he has been elected to something, even mayor of Hyannis Port. Being appointed to office isn't enough."

She inscribed a leather-bound copy of his *The Enemy Within:* "To Bobby, who made the impossible possible and changed all our lives, Jackie."

Throughout the years Bobby Kennedy remained a great help to Jackie Kennedy and the children:

"I think he is the most compassionate person I know, but probably only the closest people around him—family, friends, and those who work for him—would see that. People of a private nature are often misunderstood because they are too shy and too proud to explain themselves."

During Robert Kennedy's campaign for the Senate, Jackie met with Dorothy Schiff, publisher of the *New York Post,* whose endorsement he needed:

"He must win. He will win. He must win. Or maybe it is just because one wants it so much that one thinks that. People say he is ruthless and cold.

He isn't like the others. I think it was his place in the family, with four girls and being younger than two brothers and so much smaller. He hasn't got the graciousness they had. He is really very shy, but he has the kindest heart in the world."

ETHEL KENNEDY

Jackie was quoted as saying of Ethel, "She drops kids like the rabbits," and she once produced a caricature of Hickory Hill that depicted kids hanging out of windows and underfoot everywhere.

According to Truman Capote, Jackie referred to Ethel Kennedy as "the baby-making machine—wind her up and she becomes pregnant."

EDWARD M. KENNEDY

Wisconsin, 1959: "Ted is such a little boy in so many ways. The way he almost puffs himself up when he talks to Jack. He hero-worships him, of course. I think it was only last year that Ted started calling him Jack, and I think he first asked if he could. But there's been such a real change in him. He used to be so terribly, terribly serious all the time, but now he relaxes a little more, smiles a little more, and he's still very serious. But he's so very nice and so very intelligent."

JOAN KENNEDY

Jacqueline Kennedy liked Joan Kennedy more than the other Kennedy women did and took her under her wing: "In the beginning, Joan was so happy with Ted. Whenever we were all in Hyannis Port, you could see the pride on Ted's face when she walked in the room with her great figure and her leopard-skin outfits. If only she had realized her own strengths instead of looking at herself in comparison with the Kennedys. Why worry if you're not as good at tennis as Eunice or Ethel when men are attracted by the feminine way you play tennis? Why court Ethel's tennis elbow?"

POLITICAL WIFE

Shortly after her marriage to Jack, Jackie spoke to another Senate wife who had attempted to warn her about what she was getting into:

"My God. You told me what it would be like, but you really didn't tell me everything. You only told me half."

★

Jackie's first year of marriage was admittedly difficult:

"I was alone almost every weekend. It was all wrong. Politics was sort of my enemy, and we had no home life whatsoever."

★

The main thing for me was to do whatever my husband wanted. He couldn't—and wouldn't—be married to a woman who tried to share the spotlight with him. I thought the best thing I could do was to be a distraction. Jack lived and breathed politics all day long. If he came home to more table-thumping, how could he ever relax."

During a lull in family conversation, early in their marriage, Jack Kennedy turned to his wife and said, "A penny for your thoughts."

She replied, "But they're my thoughts, Jack, and they wouldn't be my thoughts anymore if I told them. Now would they?"

All the Kennedys looked at one another; then Joe Sr. laughed and said something about liking "a girl with a mind of her own—a girl just like us."

After Jack failed to become the vice-presidential candidate in 1956, Jackie said: "I think when something like that starts and gets momentum then you want it, yes. I know when it was over and we flew back to New York afterward there was kind of a let-down."

✱

*R*eally the way that Jack got [to be president] was all those years he'd been going around the country—it was six years of our marriage . . . of every single moment of free time going out. . . . It is that road work and it is knowing every county person or delegate that in the end got him the nomination, when he really had the least chance, I suppose, being Catholic."

✱

*A*bout Jack's long hours of devotion to politics, Jackie would further recall: "I was always coming down to breakfast in my wrapper with Caroline and there would be a couple of strange governors or labor leaders I'd never seen before, smoking cigars and eating scrambled eggs."

★

*T*here were, of course, times during those years when the demands were lessened by the joys:

"[O]ne of the most thrilling moments of my life was the time I watched Jack make his speech about the St. Lawrence Seaway."

★

*A*bout their social life in the capital city, she once said: "You always hear about Washington—all this going out and party circuit. We never did that; we didn't like it. Then Jack would be traveling a lot, so we just liked to stay home. . . . I never like to say 'enjoy the social life' because I think that sounds so trivial and frivolous. . . . Who's in? Who's out? Dinner parties . . . that silly treadmill I have no esteem for. . . . All the big receptions, the cocktail parties—forget that. I think I may have been to one in my life—or the big embassy dinners, even. I'm not sure that anything of substance is really accomplished there."

★

In 1957 Jackie and her sister, Lee, posed for a fashion spread to appear in the December issue of *Ladies' Home Journal*. Much to her husband's amusement, she was quoted in the magazine as having said: "I don't like to buy a lot of clothes and have my closets full. A suit, a good little black dress with sleeves, and a short evening dress—that's all you need for travel."

★

Jackie initially had problems grasping the reality that in politics someone could be your friend one week and your enemy the next, and vice versa. When JFK spoke agreeably about a political colleague he had previously disagreed with, his wife, dumbfounded, would say: "Why are you saying nice things about that rat? I've been hating him for three weeks now."

★

With time, however, she began to take a different view. Early in 1958, when Jack was a senator, she said: "Politics is in my blood. I know that even if

Jack changed professions, I would miss politics. It's the most exciting life imaginable—always involved with the news of the moment, meeting and working with people who are enormously alive, and every day you are caught up in something you really care about. It makes a lot of other things seem less vital. You get used to the pressure that never lets up, and you learn to live with it as a fish lives in water."

You never had the feeling that members of the opposite party were unfriendly in the Senate, either with the senators or with their wives. There was always a nice feeling. . . . You get a lot done in the Senate with bipartisan[ship]."

★

Historic Massachusetts made an impression during a campaign swing in 1958:

"I'm so glad Jack comes from Massachusetts because it's the state with the most history. Driving from one rally to another, we'd pass John Quincy

Adams's house or Harvard—or Plymouth. I think I
know every corner of Massachusetts."

✶

*I*n that 1958 Senate campaign, Jackie made her first
campaign speech, telling the Worcester Cercle
Français that public speaking in French was "not as
frightening as it would have been in English."

✶

*D*uring that same campaign, she told her stepbrother Yusha Auchincloss: "Nothing disturbs me as
much as interviewers and journalists. That's the
trouble with a life in the public eye. I've always hated
gossip-column publicity about the private lives of
public men. But if you make your living in public
office, you're the property of every taxpaying citizen.
Your whole life is an open book."

✶

*L*ook, it's a trade-off. There are positives and nega-
tives to every situation in life. You endure the bad

things, but you enjoy the good. And what incredible opportunities—the historic figures you meet and come to know, the witness to history you become, the places you would never have been able to see that now you can. One could never have such a life if one wasn't married to someone like that. If the trade-off is too painful, then you just have to remove yourself, or you have to get out of it. But if you truly love someone, well. . . ."

*E*ventually Jackie developed a more objective view of the political life and said on various occasions: "I separate politics from my private life, maybe that's why I treasure my life at home so much. In this business there're always going to be flare-ups about something. And you must somehow get so it doesn't upset you. I think I was always good at it. I can drop this curtain in my mind."

I think every woman wants to be needed, and in politics, you are."

*

I don't know anything about politics, or didn't until after I got married. Then I heard so much of it all around me all the time that I learned about politics through a kind of osmosis."

*

On the role of the political wife, Jackie had said: "You have to do what your husband wants you to do. My life revolves around my husband. His life is my life. It is up to me to make his home a haven, a refuge, to arrange it so that he can see as much of me and his children as possible—but never let the arrangements ruffle him, never let him see that it is work. I want to take such good care of my husband that, whatever he is doing, he can do better because he has me. His work is so important."

*

While delivering the eulogy at Jackie Kennedy Onassis's funeral mass, Senator Edward Kennedy recalled fondly that Jackie not only learned the ultimate lesson in politics but put it to good use during the Clintons' visit to Martha's Vineyard in 1993.

"When we were waiting for President and Mrs. Clinton to arrive, Jackie turned to me and said, 'Teddy, you go down and greet the president.' But I said, 'Maurice [Tempelsman, Jackie's late-in-life companion] is already there.' And Jackie answered, 'Teddy, *you* do it. Maurice isn't running for reelection.'"

The 1960 Presidential Campaign

I married a whirlwind. He's indestructible. People who try to keep up with him drop like flies, including me. It sounds endless and it is. The first two days [of the campaign] were the hardest—but then I got into the rhythm of it."

★

Initially in 1960, before Jack entered the presidential race: "It isn't the right time of life for us. We should be enjoying our family, traveling, having fun."

Primary campaign, Wisconsin, 1959: "We didn't talk much when we flew. Jack looked out at the farms and said you could really tell they were family farms, set all apart, all by themselves, and he made a note of it for his next speech at Rice. And he asked Ted Sorensen if he had any jokes. He was always looking for jokes."

*

Mrs. Kennedy had become friends with Mrs. John Sherman Cooper, who gave her several tips she had picked up while campaigning with her husband in Kentucky:

"I remember her telling me that she carried little cards with her, and that whenever she left a city or town, she'd write a note. She told me to do this while campaigning. . . . Right away when you leave, write a little note, 'Dear so-and-so, thank you for this or that,' because otherwise everything piles up and you forget."

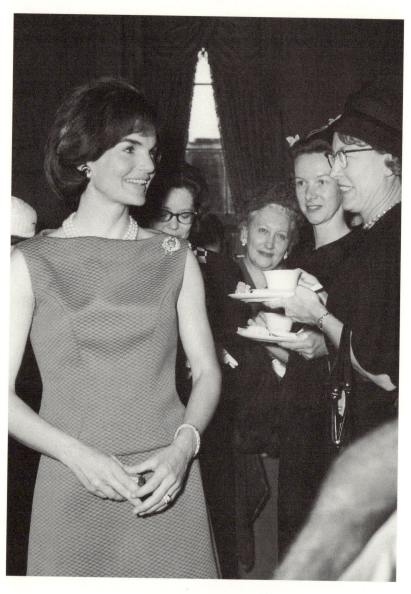

*Jackie and guests enjoy tea in the State Dining Room at the White House.
(Photo by Robert Knudson. Courtesy of the John Fitzgerald
Kennedy Library, Boston, Massachusetts.)*

★

*I*f Jack didn't run for president [in 1960], he'd be like a tiger in a cage."

★

*F*rom a television interview during the 1960 presidential campaign: "The most important thing [a political wife] needs is to really love her husband. Then, any sacrifices or adjustments she has to make are only a joy."

★

*I*n a 1960 Hyannis Port interview, she was asked if her husband was different since becoming a presidential candidate:

"I don't think Jack has changed much, I really don't. He still thinks nothing of answering the door at home when he's wearing his shorts."

★

Speaking in Georgetown about Senator Kennedy's 1960 presidential primary run against Hubert Humphrey: "When I saw his schedule . . . I told him it was silly zigzagging back and forth, and he agreed. He told me to talk it over with Bob Wallace. I did, and things were changed. That's the first time Jack told me to go ahead and do anything like that."

★

On providing literary quotes and allusions for Senator Kennedy's speeches:

"I thought of some lines from a poem I thought he ought to use, and he told me to get the rest of it. . . . I used to worry myself sick when Jack said to me that he didn't know what he was going to say in his next speech, but now, even though he still says it sometimes, it doesn't bother me because he has picked up so much more self-confidence in himself and his speechmaking that he can get up without any speech and I absolutely know he'll be all right without fumbling for thoughts or anything because he has so much in his head and he has real presence. I think it's a compliment that I listen to his speeches the way I do because he always has some fresh things to say at the beginning of each speech, things that nobody knows he was going to say. Even in the things he'd said before, the sections of speeches, he always changes them somehow so that each time it's just a little bit different."

*I*t's not easy, this traveling, but we are together and he tells me how much it helps him just for me to be there. And I try to be natural with people. I think if you aren't, then they sense it immediately."

*I*n Ashland, Wisconsin, 1959: "Jack woke me up and Steve [Smith] came in, and while he and Steve were talking about the news stories and things like that, I packed my bag and got dressed. Neither of us is very talkative so early in the morning, especially me. I don't think we said much in the car going out to the airfield. But I remember something in the car going to the airport in Ashland. I saw a crow and I told Jack we must see another crow, and I told him the jingle I learned as a little girl: 'One crow sorrow, two crows joy, three crows a girl, four a boy.' And you should have seen Jack looking for crows until he found more. He would have liked to find four crows. I guess every man wants a boy. But that was a tender thing, I thought."

✳

Campaigning in Marshfield, Wisconsin, before seventy-five people at a luncheon in the Charles Hotel: "We've been working so hard in Wisconsin, and I know that if you do see fit to support my husband, you will find you haven't misplaced your trust. In recent years he has served on the Senate Labor and Public Welfare Committee and in that capacity has done as much for workers in this country as any U.S. senator. He will continue to do everything in his power if elected president."

✳

On the campaign trail: "Everyone knows campaigning is rewarding. I have seen this country in such detail and every kind of person who lives here;

but it is the most grueling thing in the world. It was a life for which I had no preparation."

*

In all the places we campaigned—and sometimes I was so tired I practically didn't know what state we were in—those are the people who touched me the most— The poverty hit me more . . . because I just didn't realize that it existed in the U.S.—little children on rotting porches with pregnant mothers— young mothers—but all their teeth gone from bad diet."

*

[Women] are very idealistic and they respond to an idealistic person like my husband."

*

I certainly would not express any views that were not my husband's. I get all my views from him. Not because I can't make up my mind on my own, but because he would not be where he is unless he was one of the most able men in his party."

∗

The aged shouldn't be put to the awful indignity of swearing they are in need of medical care. And it's a problem for many women in their middle years. So many of them are faced with the need for taking care of their own children. College education for their children could mean leaving their parents stranded."

∗

A terrible, frightening decade is ahead. People are too complacent about this country's power. Someone has to talk to the Russians. If my country were in Jack's hands, to give the decade a start, I'd feel safe."

∗

In Spanish Harlem on the campaign trail: "My Spanish is poor but my knowledge of your history, culture, and problems is better. I can assure you if my husband is elected president you will have a real friend in the White House."

*

In response to rumors that Kennedy was really running to get the number-two slot under Adlai Stevenson: "Let Adlai get beaten alone. If you don't believe Jack, I'll cut my wrists and write an oath in blood that he'll refuse to run with Stevenson."

*

I'm not in the habit of arriving at a campaign stop before my husband, and as I don't have any speech prepared, I can tell you whom *I'm* voting for in November."

*

October 1960: "This week I made some radio tapes appealing to Puerto Ricans, Mexican Americans, Haitians, and Poles to register and vote. I am grateful to my parents for the effort they made to teach us foreign languages. All these people have contributed so much to our country's culture, it seems a proper courtesy to address them in their own tongue."

★

*I*t's so boring when we keep going to places where everybody loves Jack. . . . Going into Humphrey territory . . . makes things more fun, more challenging."

★

*Y*ou shake hundreds of hands in the afternoon and hundreds more at night. You get so tired you catch yourself laughing and crying at the same time. But you pace yourself and you get through it. You just look at it as something you have to do. You knew it would come and you knew it was worth it.

"The places blur after a while, they really do. I remember people, not faces, in a receiving line. The thing you get from these people is a sense of shyness and anxiety and shining expectancy. These women who come up to see me at a meeting, they're as shy as I am. Sometimes we just stand there smiling at each other and just don't say anything."

Asked what would happen if her husband lost the nomination: "I guess it would be like a racing-car driver who is way ahead and winning the race and then someone tells him there is no more gas for his car."

On the campaign trail: "If you can't cope with emergencies by the time you're twenty-five, you'll never be able to adapt yourself to situations."

She even charmed the manager of a Kenosha, Wisconsin, supermarket into letting her interrupt his recitation of sale items over the store's loud-speaker. Customers were astonished to hear a soft voice say: "Just keep on with your shopping while I tell you about my husband, John F. Kennedy."

She went on to talk about his career in the Navy and in Congress, spoke of how deeply he cared for his country, and ended by saying: "Please vote for him."

*

During the campaign, she had to answer criticism about her husband's religion:

"I think it's so unfair of people to be against Jack because he's a Catholic. He's such a poor Catholic. Now, if it were Bobby, I could understand it."

*

Jackie encountered just as much criticism for spending too much money on clothes:

"That's dreadfully unfair. They're beginning to snipe at me about as often as they attack Jack on Catholicism. I'm sure I spend less than Mrs. Nixon on clothes. She gets hers at Elizabeth Arden, and nothing there costs less than two hundred or three hundred dollars."

*

During the campaign, a New York trade paper contended that Mrs. Kennedy was "too chic" and claimed that she and her mother-in-law, Rose, spent $30,000 a year buying Paris clothes. She reacted angrily, saying: "I couldn't spend that much unless I wore sable underwear."

In a letter written in reply to all the unfavorable mail she was getting about her clothes and hairstyle during the presidential campaign, she said: "All the talk over what I wear and how I fix my hair has me amused, but it also puzzles me. What does my hairdo have to do with my husband's ability to be president?"

Meeting with reporters after her husband secured the nomination, Mrs. Kennedy said about her role: "I supposed I won't be able to play much part in the campaign, but I'll do what I can. I feel I should be with Jack when he's engaged in such a struggle, and if it weren't for the baby, I'd campaign even more vigorously than Mrs. Nixon. I can't be so presumptuous as to think I could have any effect on the outcome, but it would be so tragic if my husband lost by a few votes merely because I wasn't at his side and because people had met Mrs. Nixon and liked her."

*

While campaigning in Wisconsin, Mrs. Kennedy had gone off by herself to campaign among the black churches. When Senator Kennedy picked her up later in the day, he asked her how she did. She said: "Oh, I did very well. I met the loveliest minister of the loveliest black church, and he has all kinds of financial troubles. I thought it would be nice to help him out, so I gave him two hundred dollars."

JFK said, "Well, that was nice," then on second thought, added, "Goddammit, it wasn't my money, was it?"

*

October 1960: "I am not sure I share the supposed dream of American women to see their sons be president. Being president is one thing. You could not help but be proud of that. But running for office is another—an ordeal you would wish to spare sons

and husbands. You worry and wish you could dimin-
ish the strain, but of course, you cannot."

★

After some time spent campaigning: "I've never
had much of a social conscience, but now I do."

★

Pregnancy spared Mrs. Kennedy from much of the
presidential campaign:

"Thank God, I get out of those dreadful chicken
dinners. Sitting at head tables where I can't have a
cigarette and having to wear those silly corsages and
listen to some gassy old windbag drives me up the
wall. Poor Jack."

★

To Lady Bird Johnson, in Hyannis, 1960: "I feel so
totally inadequate, so totally at a loss. Here I am at
the time Jack needs me most, and I'm pregnant, and
I don't know how to do anything."

*

At home in Georgetown while pregnant, Jackie hosted women's meetings. She also wrote a syndicated newspaper column, "Campaign Wife." She discussed her potential role as first lady:

"I wouldn't put on a mask and pretend to be anything that I wasn't."

*

Writing about education and the federal government in "Campaign Wife," 1960: "Although I certainly agree that education is primarily a local responsibility . . . it does seem imperative that the federal government step in and do its share. . . . More teachers must be trained, but . . . they must be paid more so they will enter the teaching profession."

*

Guns have cost us all so very, very much. It's so sad that anyone—even the police—have to carry guns, much less ever have to use them. I wonder if there will ever be a time when the guns of the world will be nothing more than antiques—reminders to future generations of what the world was like back in the 1960s or 1970s."

*

Also excerpted from "Campaign Wife," 1960: "Though my child [Caroline] won't even begin nursery school until next year, I worry already about where to send her, all the way through high school, and sound out friends with older children about which school in our area is best for which type of child. Many of them are happy with their children's schooling—but most are not. Several days ago I read that the Prince George's County School Board in Maryland had bought trailers to be used as classrooms to reduce double sessions in elementary schools. It emphasized the urgent need all over the country for additional classrooms. My husband has been deeply concerned with these problems and has recently supported in the Senate a successful effort to pass legislation providing federal aid for school construction and for teachers' salaries."

★

During a campaign luncheon in New York, Jack and Jackie Kennedy were sitting about five seats apart, and she said: "This is the closest I've come to lunching with my husband in months. I haven't seen him since Labor Day."

★

At the end of the 1960 campaign, in Hyannis Port, to Jack, in a broadcast over network television: "Jack, I've enjoyed watching this program tonight. I only wish I could have been there with you at the end of this longest and busiest day for you—[the end of] the long road that we've traveled together since the primaries in January. The doctor wouldn't let me leave Hyannis Port tonight so I'll be here until tomorrow morning at six-thirty when I drive to Boston to join you en route. I wouldn't miss it for anything. And then we'll have you back with us at least to wait out the election returns together.

"One of my happiest memories of this campaign has been all of the people who have believed in you, who worked so hard and helped so much. I want to thank all of you who are listening tonight and tell you that we'll never be able to repay our debt to you, and that it is you that we thank tonight with gratitude."

*

Somebody mentioned it would be good politics to have her baby on the eve of the election. Her reply: "Oh, I hope not. I'd have to get up the next day to go and vote."

*

On election day, November 8, 1960, she informed Arthur Schlesinger that she had cast only one vote— for John F. Kennedy:

"It is a rare thing to be able to vote for one's husband as president of the United States, and I didn't want to dilute it by voting for anyone else."

*

One day in a campaign can age a person thirty years."

The 1960 Election

John F. Kennedy won the election against Richard Nixon, but Jackie was not able to attend all the activities owing to her pregnancy and the birth of John F. Kennedy Jr. on November 25, 1960.

*

I had been in my room for days, not getting out of bed. I guess I was just in physical and nervous exhaustion, because the month after John's birth was just the opposite of recuperation. I missed all the gala things. I always wished I could have participated more in those first shining hours with Jack, but at least I had given him our John, the son he longed for so much."

*

Before Kennedy was even inaugurated, there was a threat against his life in Palm Beach. Jackie expressed her concern: "We're nothing but sitting ducks in a shooting gallery."

★

*I*n the hospital after giving birth to John Jr., Jackie sat on the hospital sun roof. Another patient wandered by and said, "You're Mrs. Kennedy, aren't you? I recognize you from your pictures."

Mrs. Kennedy answered: "I know. That's my problem now."

★

*F*ollowing the birth of John Jr., they went to the Palm Beach home of the elder Kennedys:

"It was so crowded that I could be in the bathroom, in the tub, and then find that Pierre Salinger was holding a press conference in my bedroom."

★

[*I*t's] as though I have just turned into a piece of public property. It's really frightening to lose your anonymity at thirty-one."

★

On the loss of her much-valued privacy, she once said: "Sometimes I think you become sort of a—there ought to be a nicer word than freak, but I can't think of one."

The Inauguration

On President Kennedy's inaugural address: "I had heard it in bits and pieces many times while he was working on it in Florida. There were piles of yellow paper covered with his notes all over our bedroom floor. That day, when I heard it as a whole for the first time, it was so pure and beautiful and soaring that I knew I was hearing something great. And now I know that it will go down in history as one of the most moving speeches ever uttered—with Pericles' Funeral Oration and the Gettysburg Address."

Regarding her inauguration outfit, a beige wool coat with a sable collar and matching sable muff: "I just didn't want to wear a fur coat. I don't know why, but perhaps because women huddling on the bleachers always looked like rows of fur-bearing animals."

★

Remembering when she and Jack met at the Capitol for the first time as president and first lady: "I was so proud of Jack. There was so much I wanted to say. But I could scarcely embrace him in front of all those people, so I remember I just put my hand on his cheek and said, 'Jack, you were so wonderful!' And he was smiling in the most touching and most vulnerable way. He looked so happy."

THE NATION'S FIRST LADY

I'll be a wife and mother first, then first lady."

★

Speaking to reporters about soon becoming the first lady, she shared the following exchange with the press:

"When you are first lady, you won't be able to jump into your car and rush down to Orange Country to go foxhunting."

"You couldn't be more wrong," Jackie replied. "That is one thing I won't ever give up."

"But you'll have to make some concessions to the role, won't you?"

"Oh, I will. I'll wear hats."

✴

When JFK won the presidency and Jackie began to panic about her official duties, she told a friend: "I'll get pregnant and stay pregnant. It's the only way out."

✴

When asked on the *Today* show (September 15, 1960) what the basic duties of a first lady were, she replied: "I have always thought the main duty is to preserve the president of the United States so he can be of best service to his country, and that means running a household smoothly around him, and helping him in any way he might ask you to."

However, she never liked the title "first lady": "It always reminded me of a saddle horse."

And she was less than enthused about the traditional role of first lady: "Why should I traipse around to hospitals playing Lady Bountiful when I have so much to do around here?"

✴

A year passed and she began to warm to the role:

"I know so much more about it now. Think of this time we're living through. Both of us young, with health and two wonderful children . . . and to live through all this."

★

I didn't want to go down into coal mines or be a symbol of elegance. . . . I will never be a committee woman or a club woman. I am not a joiner. . . . Whoever lives in the White House must preserve its traditions, enhance it, and leave something of herself there. . . . I do have an official role as wife of the president, and I think every first lady should do something in this position to help the thing she cares about. I would hope that when I leave here I will have done something to help, for instance, the arts in which I am so interested. . . . People seem so interested in whatever the first family likes. That is where I think one can lead. One doesn't know whether one leads in the right direction or not, but one hopes one does."

★

To the public she said: "My husband loves a challenge, and I do too. I would hope that when I leave here I will have done something to help."

★

On another occasion she confided: "I seem so mercilessly exposed and don't know how to cope with it."

★

What do you suppose they want me to be? I've always been the same person. . . . I always felt I was myself, but with so many reporters watching, listening, how can anyone not seem like someone you're not?"

★

Mrs. Kennedy also once said she personally felt "the most affinity" for Mrs. Harry Truman:

"She brought a daughter to the White House at

a most difficult age and managed to keep her from being spoiled so that she has made a happy marriage with lovely children of her own. Mrs. Truman kept her family close together in spite of White House demands, and that is the hardest thing to do."

★

Sometimes I get furious with myself thinking about all the energy I wasted worrying about what life would be like in the White House. We had such a wonderful home in Georgetown. You'd come in at night and find the fire going, and people talking, and you didn't stay up late. And my fears were that we wouldn't have this anymore in the White House. But that's been the most wonderful side of it. You can talk when he comes home at night. It's better than during the campaign—you don't just dump your bags and go off again. And that's been wonderful for the children. Sometimes they even have lunch with Jack—if you'd told me that would happen, I'd never have believed it.

"But I should have realized, because, after all, the one thing that happens to a president is that his ties with the outside world are cut. And the only people you really have are each other. I should think that if people weren't happily married, the White House would really finish it."

★

We never talked of serious things. I guess because Jack has always told me the one thing a busy man doesn't want to talk about at the end of the day is whether the Geneva Convention will be successful or what settlement could be made in Kashmir or anything like that. He didn't tell me those things. He wanted me as a wife and seldom brought home his working problems—except once in a while the serious ones."

★

At her first official press conference, Jackie spoke of her priorities:

"I have no desire to influence fashions. That is at the bottom of any list."

What was at the top?

"Jack."

While the president left matters of the home to his wife, she would sometimes cajole his input. She once explained to an interviewer:

"When I start to ask him silly little insignificant questions about whether Caroline should appear at some reception, or whether I should wear a short or long dress, he just snaps his fingers and says, 'That's your province.' And I say, 'Yes, but you're the great decision-maker. Why should everyone but me get the benefit of your decisions?'"

When her husband complained about her spending, she responded: "I have to dress well, Jack, so I won't embarrass you. As a public figure, you'd be humiliated if I was photographed in some saggy old housedress. Everyone would say your wife is a slob and refuse to vote for you."

On yet another occasion, she quipped: "The president seems more concerned these days with my budget than with the budget of the United States."

She told the chief usher in 1961: "I want my husband to be able to leave the office, even for a few hours. I want to surround him with bright people who can hold his interest and divert his mind from what's going on over there!"

Being in the White House does make friendships difficult. Nobody feels the same. Jack's even more isolated than I, so I do try to have a few friends for dinner as often as possible. Mostly it turns out to be the Charlie Bartletts or Bill Walton or someone we know really well, because I hate to call and have people feel they have to come."

In 1962, after a year as first lady, Jackie reflected: "When we first moved into the White House, I was too exhausted at the end of the day to do anything but the state dinners and things we just had to do. Now I'm better organized. I work from eight o'clock in the morning until noon at my desk and try to save lunch and as much of the afternoon as I can for the children."

Jackie also held the conviction that reporters who were personal friends of theirs shouldn't ask JFK tough questions. Accompanying him to an appearance on *Face the Nation,* she slipped notes onto the desks of journalists she knew, saying, "Don't ask Jack mean questions."

Responding to the seemingly endless stream of interview requests she received proved tiresome as well. She wrote to one editor:

"The thing is—we have done a whole rash of stories lately—four this summer and another coming up [an interview with Richard Rovere for *McCall's*]. The ones we've done are *Life, Look, Ladies' Home Journal,* and *Redbook.* I am so tired of all the hard work and confusion that goes into a story—especially one with pictures, and feel pretty stale right now. . . . Couldn't you just use some of Jacques Lowe's old pictures—he just took some recent ones of us—trying to get something Jack could use as a Christmas card—please no little photographic essays, Jacques Lowe and I have been through about three sessions like that together—changing clothes, fixing lights, driving to find nice scenery, trying to make the baby smile— I'm sure he wants to avoid it as much as I do!"

★

Turning down yet another request for a story, she wrote: "I wish I could either tell you that I would love to do it—or had just been run over by a bus— and couldn't pose for a month. They are marvelous articles—but if you won't be too angry—I think I would just as soon not do one—revealing my few tragic beauty secrets and disorganized wardrobe!"

★

Mrs. Kennedy also wanted the press to be ejected once the White House state dinners got under way:

"That is when they ask everyone questions and I don't think it is dignified to have them around. It always makes me feel like some social-climbing hostess. Their notebooks also bother me, but I think they should be made to wear big badges and be whisked out of there once we all sit down to dinner."

★

In another memo she suggested that members of the press "be permitted to attend important receptions but be kept out of sight behind the pillars and potted palms. They are too intrusive. They surround our guests and monopolize them. Nobody could get near John Glenn the other night. Also, the minute the photographers have finished shooting, they are to be ushered out the front door, so the Marine Band can strike up 'Hail to the Chief.'"

★

By order of the president, photographers were barred from taking pictures of her smoking:

"He's such a bear about my smoking that I started encouraging him to have a cigar after dinner. That way he doesn't complain so much about my cigarettes."

✶

During Jackie's trip to India in 1962, the international press criticized her for wearing high fashion in a country overrun with poverty. She subsequently instructed her press attaché to refrain from providing further fashion information:

"If you say anything, tell them it's secondhand and that I bought everything at the Ritz Thrift Shop."

✶

In Venezuela she said: "No parents could be happy until they have the possibility of jobs and education for their children. This must be for all and not just for a fortunate few."

The President and First Lady with the emperor of the Ethiopia, Haile Selassie. (Photo by Cecil W. Stoughton. Courtesy of the John Kennedy Library, Boston, Massachusetts.)

*

In Colombia: "I know that we share the desire of bringing to all the peoples of the hemisphere a better life for themselves, and know that you agree that the good things in life—education, housing, and employment—should be within the reach of all and not just a few blessed by fortune."

*

After her trip to India and Pakistan: "Jack's always so proud of me when I do something like this, but I can't stand being out in front. I know it sounds trite, but what I really want is to be behind him and to be a good wife and mother. I have no desire to be a public personality on my own."

*

Because of the increased speculation over the cost of her clothes, Jackie wanted designer Oleg Cassini to consult with her in advance regarding the release of any information about her wardrobe:

"I don't want to seem to be buying too much. . . . There just may be a few things we won't tell them about! But if I look impeccable the next four years, everyone will know it is you."

Later she wrote to Cassini that she wanted to be dressed as if "Jack were President of France." She knew she was "so much more of fashion interest than other first ladies," but didn't want to be "plagued by fashion stories of a sensational nature." She didn't want to be "the Marie Antoinette or Empress Josephine of the 1960s" or to "seem to be buying too much." She requested that he "make sure no one has exactly the same dress as I do. . . . I want all mine to be original and no fat little women hopping around in the same gown."

When she was first lady, Jackie read that her sister was supposedly more elegantly dressed. In response, she wrote to one of her "fashion spies" in Paris:

"What I really appreciate most of all is your letting me know before Lee about the treasures. Please always do that—now that she knows you are my 'scout,' she is slipping in there before me. So this fall, do let me know about the prettiest things first."

Though Jackie was aware of her husband's reported extramarital escapades while he was president, she never spoke of or confronted him about his affairs, except once when she discovered a woman's undergarment beneath a pillow in their White House bedroom. Holding it between thumb and forefinger, she brought it to Jack and said:

"Would you please shop around and see who this belongs to? It's not my size."

When Jacqueline took a French photographer on a tour of the White House, she opened an office door and told the photographer in French:

"And this is a young lady who is supposed to be sleeping with my husband."

★

There is a story that Marilyn Monroe once called Mrs. Kennedy in the White House to tell her she was having an affair with the president. Supposedly Marilyn asked what Jackie thought about the possibility of Marilyn marrying Kennedy, and the response was:

"Marilyn, you'll marry Jack, that's great, and you'll move into the White House and you'll assume the responsibilities of first lady, and I'll move out and you'll have all the problems."

★

Mrs. Kennedy is said to have gone into the marriage with her eyes open. She told a friend: "I don't think there are many men who are faithful to their wives. Men are such a combination of good and evil."

★

She also told Joan Kennedy: "Kennedy men are like that. They'll go after anything in skirts. It doesn't mean a thing."

*

My husband was a romantic, although he didn't like people to know that."

*

I know my husband was devoted to me. I know he was proud of me. It took a very long time for us to work everything out, but we did, and we were about to have a real life together. I was going to campaign with him. I know I held a very special place for him—a unique place. . . . Jack was something special, and I know he saw something special in me too. . . . The three years we spent in the White House were really the happiest time for us."

*

John F. Kennedy was reportedly jealous of Mrs. Kennedy's male friends. She was aware of the gossip and told Robin Douglas-Home: "What can I do? I have dinner with someone, dance with someone for more than one dance, stay with someone, get photographed with someone without Jack—and then everyone automatically says, 'Oh, he must be her new lover.' How can you beat that?"

In a memo to her social secretary dated January 11, 1963, she wrote: "I'm taking the veil. I've had it with being First Lady all the time and now I'm going to give more attention to my children. I want you to cut off *all* outside activity—whether it's a glass of sherry with a poet or coffee with a king. No more art gallery dedications—no nothing—unless absolutely necessary."

She was expecting her third child in the fall and wanted to keep the pregnancy out of the news until April, when the White House would make an official announcement. On trying to keep her pregnancy secret, she told the president:

"I don't know how I'm going to keep this a secret until then. I have an uncanny ability to sense when someone is pregnant, and I just know someone like me is going to find out about it."

✴

Off the record, she told journalist Charlotte Curtis, her former dormitory mate at Vassar: "The White House is such an artificial environment. It's a snake pit. If I don't take care of myself, I'll go mad."

✴

I just wanted to save some normal life for Jack and the children and for me. My first fight was to fight for a sane life for my babies and their father."

The White House

My mother brought me to Washington one Easter when I was eleven. That was the first time I saw the White House. From the outside I remember the feeling of the place. But inside, all I remember is shuffling through. There wasn't even a booklet you could buy. Mount Vernon and the National Gallery of Art and the FBI make a far greater impression. I remember the FBI especially because they finger-printed me."

After her tour of the White House with Mrs. Eisenhower, who preceded her as the first lady, Jackie exclaimed: "Oh, God. It's the worst place in the world. So cold and dreary. A dungeon like the Lubyanka. It looks like it's been furnished by dis-count stores. I've never seen anything like it. I can't bear the thought of moving in. I hate it, hate it, hate it."

On her visit to the White House with Mrs. Eisenhower, Jackie told Letitia Baldrige, her new social secretary, that the place looked like "a hotel that had been decorated by a wholesale furniture store during a January clearance."

*

Complaining to *Life*'s Hugh Sidey: "[It's] like a hotel. Everywhere I look there is somebody standing around or walking down a hall."

*

I felt like a moth banging on the windowpane when I first moved into this place. It was terrible. You couldn't even open the windows in the rooms, because they hadn't been opened for years. The shades you pulled down at night were so enormous that they had pulleys and ropes. When we tried the fireplaces, they smoked because they hadn't ever been used. Sometimes I wondered, 'How are we going to live as a family in this enormous place?' I'm afraid it will always be a little impossible for the people who live there. It's an office building."

*

*I*n a 1961 article in *Life* by Hugh Sidey, she was quoted: "The minute I knew that Jack was going to run for president, I knew the White House would be one of my main projects if he won."

★

*L*ater in the same interview: "Like any president's wife I'm here for only a brief time. And before everything slips away, before every link with the past is gone I want to do this. I want to find all the people who are still here who know about the White House, were intimate with it—the nephews, the sons, the great-grandchildren, the people who are still living and remember things about the White House.

"It has been fascinating to go through the building with Mrs. Nicholas Longworth, who was Theodore Roosevelt's daughter, and with Franklin D. Roosevelt Jr. and President Truman, and hear them tell where things had been placed in their day."

★

I just think that everything in the White House should be the best, the entertainment that's given here, and if it's an American company that you can help, I like to do that. If it's not, just as long as it's the best."

★

I think the White House should show the wonderful heritage that this country has. We had such a wonderful flowering in the late eighteenth century. And the restoration is so fascinating—every day you see a letter that has come in from the great-great-grandson of a president. It was such a surprise to come here and find so little that had association and memory. I'd feel terribly if I lived here for four years and hadn't done anything for the house."

★

*B*efore her efforts to restore the White House she had said, "It looks like a house where nothing has taken place. There is no trace of the past."

★

*S*he often wrote memos to White House staff, this one to J. B. West, the chief usher:

"If there's anything I can't stand, it's the Victorian mirrors—they're hideous—off to the dungeons with them. Have them removed and relegated to the junk heap."

★

The first spring he was in office, President Kennedy asked Bunny Mellon, an expert horticulturist, to redesign the scraggly flower patch just outside of his office window. Jackie praised the job she did:

"It was absolutely atrocious before Bunny took over. Now it's magnificent. . . . The beauty of it seems to affect even hard-bitten reporters who come there just to watch what is going on."

★

Referring to the private living quarters at the White House, Jackie also told her decorator, Sister Parrish: "Let's have lots of chintz and gay up this old dump."

"I never want a house," she also said, "where you have to say to the children, 'Don't touch.'"

★

Commenting in 1961 on the second-story Oval Room: "This is a beautiful room. I love it most. There is this magnificent view. It means something to the man who stands there and sees it—after all he's done to get there."

★

She relished the great view down toward the Mall from the Truman balcony. She told a visitor, sweeping her arm from the Washington Monument to the Jefferson Memorial: "This is what it is all about. This is what these men fight so hard for."

★

Feeling she was being hounded by sightseers, she asked for more shrubbery to be planted by the White House fence to obscure the view from the street:

"I'm sick and tired of starring in everybody's home movies."

★

Mrs. Kennedy expounded on her policy of secrecy in a memo to her press secretary Pam Turnure, shortly after moving into the White House:

"I feel strongly that publicity in this era has gotten completely out of hand—and you must really protect the privacy of me and my children—but not offend [the press]. . . . My press relations will be minimum information given with maximum politeness. . . . I won't give any interviews—pose for any photographs, etc.—for the next four years."

★

To J. B. West, the chief usher, after demanding to see the White House bomb shelter, which also served as a command post for the Signal Corps and was filled with people working: "How amazing! I didn't expect to find so much *humanity!* I thought it would be a great big room that we could use as an indoor recreation room for the children. I even had plans for a basketball court in there!"

*

After the elegance of their visit to France, she decided to have a candlelight dinner on the lawn of Mount Vernon for their next state dinner. Aware of how complicated this would be, she said to the White House usher, "I suppose you're going to jump off the White House roof tomorrow?"

"No," replied J. B. West, "not until the day after the dinner."

*

Old Kennedy pal LeMoyne Billings, JFK's roommate from Choate, visited the White House every weekend. The chief usher always informed Mrs. Kennedy of his arrival:

"Oh, Mr. West. He's been a houseguest of mine every weekend since I've been married."

*

To J. B. West, after calling him urgently to the White House on a Sunday, his day off: "There's something brewing that might turn out to be a big catastrophe—which means that we may have to cancel the dinner and dance for the Jaipurs [the maharaja and maharani of Jaipur] Tuesday night. Could you please handle the cancellation for me. This is all very secret."

The next evening President Kennedy went on television to inform the country of the Cuban Missile Crisis.

★

Once Soviet ambassador Mikhail A. Menshikov brought a dog, named Pushinka to the Oval Office for the first lady. The puppy was one of the Soviet space dogs. When the president queried Mrs. Kennedy about it, she replied:

"I'm afraid I asked Mr. Khrushchev for it in Vienna. I was just running out of things to say."

★

To a visitor: "Look at that Lincoln cake plate. I wonder if there is enough china here to set nine places for tonight. Senator [Albert] Gore [future vice president Al Gore's father] would love to eat off Lincoln's plates."

And that night, Senator Gore did.

✶

During the televised White House tour in the Lincoln Room, Charles Collingwood asked if Jackie spent a great deal of time there:

"We did in the beginning. It was where we lived when we first came here when our rooms at the end of the hall were being painted. I loved living in this room. It's on the sunny side of the house, and one of Andrew Jackson's magnolia trees is right outside the window."

✶

Collingwood later asked if all the pieces in the Lincoln Room were from Lincoln's time.

"Yes, they are. The most famous one, of course, is the Lincoln bed. Every president seemed to love it. Theodore Roosevelt slept in it. So did Calvin Coolidge. It's probably the most famous piece of furniture in the White House. It was bought by Mrs.

Lincoln along with the dressing bureaus and chair and this table. She bought a lot of furniture for this house. She made her husband rather cross because he thought she spent too much money."

★

Collingwood pointed out the famous Gilbert Stuart portrait of George Washington. Jackie commented:

"That's the oldest thing in the White House. The only thing that was here since the beginning. A rather interesting precedent was set when that picture was painted. A commission was given to the finest living artist of the day to paint the president, and later the government bought it for the White House. I often wish they'd followed that because so many pictures of later presidents are by really inferior artists."

★

As they toured the Red Room, they discussed the art on the walls. Jackie proclaimed:

"I feel so strongly that the White House have as fine a collection of American pictures as possible. It's so important, the setting in which the presidency is presented to the world, to foreign visitors. And American people should be proud of it. We had such a great civilization, yet so many foreigners don't realize it. This little table, for instance; it's by Lannuier, a French cabinetmaker who came to America. Not many people know of him. But he was just as good as Duncan Phyfe or as the great French cabinetmakers. All the things we did so well—pictures, furniture—I think that this house should be the place where you can see them best."

Sometimes I used to stop and think about it all. I wondered, 'How are we going to live as a family in this enormous place?' I would go and sit in the Lincoln Room. It was the one room in the White House with a link to the past. It gave me great comfort. I love the Lincoln Room. Even though it isn't really Lincoln's bedroom, it has his things in it. When you see that great bed, it looks like a cathedral. To touch something I knew he had touched was a real link with him. The kind of peace I felt in that room was what you feel when going to church. I used to sit in the Lincoln Room and I could really feel his strength. I'd sort of be talking with him. Jefferson is the president with whom I have the most affinity. But Lincoln is the one I love."

The White House Restoration

I don't understand it. Jack will spend any amount of money to buy votes, but he balks at investing a thousand dollars in a beautiful painting."

★

Although President Truman had had the White House repaired structurally, the interior needed work by the time the Kennedys moved in. Mrs. Kennedy resolved to restore the White House to show it at its best:

"I want to make this into a grand house."

★

She spoke to writer Hugh Sidey of *Life*, who was doing an article on the restoration of the White House: "All these people come to see the White

House, and they see practically nothing that dates back before 1948. Every boy who comes here should see things that develop his sense of history. For the girls, the house should look beautiful and lived-in. They should see what a fire in the fireplace and pretty flowers can do for a house; the White House rooms should give them a sense of all that. Everything in the White House must have a reason for being there. It would be sacrilege merely to 'redecorate' it—a word I hate. It must be restored— and that has nothing to do with decoration. That is a question of scholarship."

When I first moved into the White House, I thought, I wish I could be married to Thomas Jefferson, because he would know best what should be done to it. But then I thought, no, presidents' wives have an obligation to contribute something, so this will be the thing I will work hardest at myself.

"How could I help wanting to do it? I don't know. . . . Is it a reverence for beauty or for history? I guess both. I've always cared. My best friends are people who care. I don't know. . . . When you read

Proust or listen to Jack talk about history or go to Mount Vernon, you understand. I feel strongly about the children who come here. When I think about my own son and how to make him turn out like his father, I think of Jack's great sense of history."

★

When she was informed that the cost of the White House renovation could create problems, Jackie devised a plan. She invited attorney Clark Clifford to the White House for lunch to get his advice on her idea.

"How many people go through the White House every year?" she asked.

Clifford didn't know, but suggested between one and two million visitors, and asked why she wanted to know.

"Before I answer your questions, you answer mine. Do any of these people leave money at the White House?"

"No. The White House is public property. People don't pay to go on the tour. Why should they?"

"They shouldn't," said Mrs. Kennedy. "But we should make available something tangible that they can buy at the White House and take away with them as a memento. We could use the money because, in effect, my goal is to make the White House 'the First House in the Land.'"

Clifford said he'd think about it.

"Don't let's think about it. Let's do something about it," she replied. "I have several ideas. One of them is to sell postcards, not the usual kind but painting postcards of the various staterooms, something the children can take home and paint over. If that's not possible, I want to put together a White House guidebook, a book with eloquent words and beautiful pictures, the kind of publication the *National Geographic* puts out for its members, but not as corny. We'll sell it for a dollar. People who go through this place in fifteen minutes can't possibly tell you what they've seen. This will remind them, and it will help pay for the renovation project. In fact it can be reprinted every time there's a new administration with material included on the ongoing administration."

★

Mrs. Kennedy eventually established the White House Historical Association to publish the guidebook to sell to visitors. Published with the assistance of the National Geographic Society, it has sold 8 million copies since 1962, with the proceeds used to benefit the White House. Mrs. Kennedy edited the book. The first one was presented to the Kennedys on June 28, 1962.

Jacqueline Kennedy was overheard saying to Chief Usher West:

"Now, J.B., I want it understood that everyone has to pay a dollar, even Ethel."

When I think of all the schoolchildren who come here, I think there should be flowers when there can be, and fires going and the pictures, to make it look rather like a home and not so frightening."

*

She spoke to Hugh Sidey of *Life*:

"I hope the Smithsonian will also maintain a permanent curator at the White House to see that things are properly cared for. For example, the famous Healy portrait of Lincoln in the State Dining Room has a damaged spot that measures eight inches across. Many other presidential portraits are in disrepair. We asked for estimates to restore pictures and frames and the total came to fifty-five thousand dollars. How can we ask Congress to appropriate that much when in these days the money is needed for so many things?

"The White House belongs to the American people. A curator would take care that it is preserved for them."

After rummaging through two basements chock-full of White House antiquities, she said: "I had a backache every day for three months, but it was a new mystery story every day."

The White House is an eighteenth- and nineteenth-century house and should be kept as a period house. Whatever one does, one does gradually, to make a house a more lived-in house with beautiful things of its period. I would write fifty letters to fifty museum curators if I could bring Andrew Jackson's inkwell home."

Shocked by the poverty she saw while campaigning in West Virginia, she ordered champagne glasses for the White House from the Morgantown Glassware guide.

"That glassware was advertised and sold everywhere as the White House wineglass—it only cost something like six dollars a dozen, but I didn't mind that at all, as I thought it was nice to help West Virginia and nice that people should see that those simple glasses were pretty enough for the White House."

Mrs. Kennedy went to great lengths to obtain items for the White House restoration project. She wanted a $10,000 Oriental rug for the president's private dining room. She wrote a memo to the chief usher regarding attempted overcharges by antique and rug dealers:

"I so like the rug but we are short of dollars and I am ENRAGED at everyone trying to gyp the White House. Tell him if he gives it he can get a tax donation and photo in our book—if not—good-bye!"

The dealer donated the rug.

Newspaper and magazine publisher Walter H. Annenberg owned a portrait of Ben Franklin that Mrs. Kennedy wanted for the White House. Clark Clifford tells how she telephoned Annenberg and related her plans for making the White House a national monument. She alluded to his priceless art collection.

"I've been told that you have a magnificent portrait of Ben Franklin by David Martin," she said.

Mr. Annenberg began to get the message.

"You, Mr. Annenberg, are the first citizen of Philadelphia. And in his day, Benjamin Franklin was the first citizen of Philadelphia. And that's why, Mr. Annenberg, I thought of you. Do you think a great

Philadelphia citizen would give the White House a portrait of another great Philadelphia citizen?"

Although the painting had cost him over $250,000, he made the donation.

*

In 1963, after her husband was assassinated and before she left the White House, she handed out mementos to friends and members of the staff. To Godfrey McHugh she said:

"First I didn't want in, now I can't seem to leave."

THE ASSASSINATION

My husband never made a sound. He had this sort of quizzical look on his face and his hand was up. I remember thinking he just looked as if he had a slight headache. And then he put his hand to his forehead and fell into my lap."

*

Jackie told her friends in early November 1963 that she dreaded the trip to Dallas:

"Jack knows I hate that sort of thing. But all he said to me was, 'I'd love you to come with me, but only if you really want to come. You would be a great help to me. But if you don't want to, I will quite understand.' So now I'm quite firm in my decision to go to Texas, even though I know I'll hate every minute of it. But if he wants me there, then that's all that matters. It's a tiny sacrifice on my part for something he feels is very important to him."

★

*A*bout the trip to Dallas:

"[Whether JFK would drop LBJ as his running mate in 1964] would be brought up every now and then and was rather annoying. I don't think he had any intention of dropping Vice President Johnson. . . . The point of the trip [was] to heal everything, to get everybody to ride in the same car. . . . I know [John] had always thought [Governor John] Connally was very intelligent . . . [and] Connally wanted to show that he was independent and could run on his own. . . . [P]art of the trouble of the trip was him trying to show that he had his own constituency."

★

*T*hree times in Texas we were greeted with bouquets of yellow roses of Texas. Only in Dallas they gave red roses. I remember thinking—how funny, red roses for me."

★

There were such big crowds of such waving, nice, happy people. I certainly did have a feeling it was going well."

★

After the assassination: "I should not have allowed him to come here. I didn't want him to come here. And he didn't want to come here. Why on earth did they make him come here? Oh, we had so *many* good times."

★

In the aftermath, she was offered a sedative:
"I don't want a sedative, I want to be with my husband when he dies."

★

Following the assassination, one of the doctors at Parkland Memorial Hospital in Dallas urged her to leave. She said:

"Do you think seeing the coffin can upset me, Doctor? I've seen my husband die, shot in my arms. His blood is all over me. How can I see anything worse than I've seen?"

★

Asked if she wanted to wipe the blood from her clothes: "Absolutely not. I want the world to see what Dallas has done to my husband."

★

Mrs. Kennedy was determined to spend the night at Bethesda Naval Hospital if necessary: "I'm not leaving here till Jack goes, but I won't cry till it's all over."

★

At Bethesda, Mrs. Kennedy spoke to her mother: "He didn't even have the satisfaction of being killed for civil rights. It had to be some silly little communist. It even robs his death of any meaning."

Mrs. Hugh Auchincloss mentioned that the children were safe at her house. Jackie had sent no message to have them taken there.

"Mummy," she said, "my God, those poor children. Their lives shouldn't be disrupted, now of all times." She thought about it. "Tell Maud Shaw to bring them back and put them to bed."

She told the story over and over again for the next few days after the assassination. To the White House chief usher, she said: "To think that I very nearly didn't go. Oh, Mr. West, what if I'd been here—out riding in Virginia or somewhere—thank God I went with him."

Among the many times she showed her concern for the feelings of others were the times of tragedy. On the day after the assassination, she wanted to contact the widow of J. D. Tippitt, the murdered Dallas police officer:

"What the poor woman must be going through."

*

At Bethesda, to Evelyn Lincoln, President Kennedy's secretary: "It's getting late and I'm going to be here for a while, so why don't you go home and try to get some rest? You hold up for the next few days, and then we'll all collapse."

*

Prior to President Kennedy's funeral procession, she reportedly told the staff: "Please be strong. In two or three days we'll all collapse."

*

Pierre Salinger, Kennedy's press secretary, remembers that at the time of the assassination, he had flown all night to get back from an official trip. When he arrived, a Mass was going on in the East Room, and Mrs. Kennedy was there. She came up to him afterward and put her arm around him and said:

"You know, Pierre, you've had a terrible day. Why don't you spend the night here in the White House."

★

After the assassination, Jackie was asked by Kenneth O'Donnell if she wanted to watch Lyndon Johnson take the oath of office. She said:

"I think I ought to. In the light of history, it will be better if I was there."

★

Aboard Air Force One on the way back to Washington, she turned to the new first lady:

"Oh, Lady Bird, it's good that we've always liked you two so much. Oh, what if I had not been there? I'm so glad I was there."

Jackie with John Jr. and Caroline leaving the Capitol after a memorial service for the slain president. Behind the First Lady is Robert F. Kennedy and Patricia Kennedy Lawford. (Photo by Abbie Rowe. Courtesy of the John Fitzgerald Kennedy Library, Boston, Massachusetts.)

★

Mrs. Kennedy is quoted as having said to Mac
Kilduff, assistant White House press secretary, on
the plane coming back to the White House after the
assassination:

"You make sure, Mac—you go and tell the president—don't let Lyndon Johnson say that I sat with
him and Lady Bird and they comforted me all during the trip. You say that I came back here and sat
with Jack."

The Funeral

After the assassination, she focused on the funeral:

"I don't want any undertakers. I want everything done by the Navy."

★

At one point, she turned to Angier Biddle Duke,
chief of protocol, and said: "Find out how Lincoln
was buried."

★

*B*efore the funeral, she wrote a letter to her husband and told Caroline: "You must write a letter to Daddy now and tell him how much you love him."

John F. Kennedy Jr., too young to write, scribbled on Caroline's letter. The letters were put in Kennedy's coffin.

★

*A*t President Kennedy's funeral, following the eulogies, Mrs. Kennedy led her daughter to the catafalque. Before the cameras, she whispered to Caroline:

"We're going to say good-bye to Daddy, and we're going to kiss him good-bye and tell Daddy how much we love him and how much we'll always miss him."

The family's official literary eulogizer later recorded:

Mother and daughter moved forward, the widow gracefully, the child watching carefully to do just as she did. Jacqueline Kennedy knelt. Caroline knelt. "You know. You just kiss," whispered Mrs.

Kennedy. Eyes closed, they leaned over to brush their lips against the flag. Caroline's small gloved hand crept underneath, to be nearer, and in that single instant an entire nation was brought to its knees. The audience in the rotunda, the national audience, those who until now had been immune, those who had endured everything else were stricken in a fraction of a second. A chord deep in the hearts of men had been touched. . . . Still clutching Caroline, she rose and stepped toward the door with simple majesty. The others stumbled after her.

★

The march to St. Matthew's Cathedral had been Mrs. Kennedy's idea.

Someone had asked, "What if it rains?"

Her reply:

"Then we'll march under umbrellas."

★

At the state funeral, as the coffin was being lashed to the caisson, Mrs. Kennedy whispered to her son,

"John, you can salute Daddy now and say good-bye to him."

The unforgettable picture of young John, his hand raised in a smart salute, wrenched hearts all over the world.

Shortly after they entered the White House, she had asked her husband where they would be buried.

He had told her, "Hyannis, I guess. We'll all be there." She had said, "Well, I don't think you should be buried in Hyannis. I think you should be buried in Arlington. You just belong to all the country."

Three years later, of course, he was buried in Arlington.

When choosing the spot where President Kennedy would be buried in Arlington National Cemetery, she remarked:

"The place is so beautiful that I could stay here forever."

The Aftermath

The day after the funeral, she wrote in her own hand from the White House to President Johnson:

> Dear Mr. President:
>
> Thank you for walking yesterday—behind Jack. You did not have to do that. I am sure many people forbade you to take such a risk, but you did it anyway. . . .
>
> Thank you for your letters to my children. What those letters will mean to them later you can imagine. The touching thing is they've always loved you so much. They were most moved to have a letter from you now, and most of all, Mr. President, thank you for the way you've always treated me, the way you and Lady Bird have always been to us before when Jack was alive, and now as president. I think the relationship of the president and vice presidential families could be a rather strained one. From the history I've been reading ever since I came to the White House, I gather it often was in the past, but you were Jack's right arm, and I always thought the greatest act of a gentleman that I had seen on this earth was how

you, the Majority Leader, when he came to the
Senate, was just another little freshman who
looked up to you and took orders from you. To
then serve as vice president to a man who had
served under you and been taught by you. But
more than that, we were friends, all four of us. All
you did for me as a friend and the happy times we
had. I always thought, way before the nomina-
tions, that Lady Bird should be the first lady, but
I don't need to tell you here what I think of her
qualities, her extraordinary grace of character, her
willingness to assume every burden. She assumed
so many from me, and I love her very much, and I
love your two daughters, Lynda Bird most because
I knew her the best, and we first met when nei-
ther of us could get a seat to hear President
Eisenhower's state of the union message, and some-
one found us a place on one of the steps in the
aisle where we sat together. It was strange last
night. I was wandering through this house. This is
the night after the funeral. There in the East
Room, I had framed the page we had all signed,
you, Senator [Everett] Dirksen, [Senator] Mike
Mansfield. Underneath, I had written the day the
vice president brought the East Room chandelier

back from the Capitol. Then in the library, I showed Bobby the Lincoln record book you gave. You see all you gave, and now you are called on to give so much more. You are the first president to sit in it as it looks today. Jack always wanted the red rug, and I had curtains designed for it that I thought were as dignified as they could be for a President's office. Late last night, a moving man asked me if I wanted Jack's ship pictures left on the wall for you. They were clearing the office for you. I said no because I remembered all the fun Jack had those first days hanging the pictures of things he liked, setting out his collection of whale's teeth, but of course they were there only for you to ask for them if the walls looked too bare. I thought you would want to put things from Texas in it. I picture some gleaming longhorns. I hope you put them somewhere. It cannot be very much help to you your first day in office to hear children out on the lawn at recess. It is one more example of your kindness that you let them stay. I promise they will be gone soon. Thank you Mr. President.

Respectfully, Jackie

★

When Mrs. Kennedy was leaving the White House after the assassination, she took Lady Bird Johnson on a complete tour and gave her some counsel:

"Don't be frightened of this house—some of the happiest years of my marriage have been spent here. You will be happy here."

Lady Bird said Jackie repeated that over and over, "as though she were trying to reassure me."

★

In her oral history provided to the Lyndon Baines Johnson Library, Mrs. Kennedy recalled making several requests of President Johnson:

"I suppose one was in a state of shock, packing up. But President Johnson made you feel that you and the children [could stay], a great courtesy to a woman in distress. It's funny what you do in a state of shock. I remember going over to the Oval Office to ask him for two things. One was to name the Space Center in Florida 'Cape Kennedy.' Now that I think back on it, that was so wrong. If I'd known

Cape Canaveral was the name from the time of Columbus, it would be the last thing Jack would have wanted. . . .

"And the other one, which is so trivial, was: there were plans for the renovation of Washington and there was this commission, and I thought it might come to an end. I asked President Johnson if he'd be nice enough to receive the commission and sort of give approval to the work they were doing, and he did. It was one of the first things he did."

★

After Dallas, Mrs. Kennedy worked hard to polish President Kennedy's image. She told a family friend, journalist Charles Bartlett: "Bobby gets me to put on my widow's weeds and go down to [LBJ's] office and ask for tremendous things."

★

In a handwritten letter to Nikita Khrushchev dated December 1, 1963, she wrote:

Dear Mr. Chairman President:

I would like to thank you for sending Mr. Mikoyan as your representative to my husband's funeral.

He looked so upset when he came through the line and I was very moved.

I tried to give him a message for you that day—but as it was such a terrible day for me, I do not know if my words came out as I meant them to. . . .

You and he were adversaries, but you were allied in a determination that the world should not be blown up. You respected each other and could deal with each other. I know that President Johnson will make every effort to establish the same relationship with you.

The danger that troubled my husband was that war might not be started so much by the big men as by the little ones.

While big men know the needs for self-control and restraint—little men are sometimes moved more by fear and pride. If only in the future the big men can continue to make the little ones sit down and talk, before they start to fight. . . .

I send this letter because I know so deeply

of the importance of the relationship between you and my husband, and also because of your kindness, and that of Mrs. Khrushchev in Vienna.

I read that she had tears in her eyes when she left the American Embassy in Moscow, after signing the book of mourning. Please thank her for that.

*Sincerely,
Jacqueline Kennedy*

✶

To the Venezuelan president: "He feared for your life and for the whole future of Latin America if anything happened to you. How strange it is—I never thought anything like this could happen to him."

✶

In a letter to Richard Nixon: "You two young men—colleagues in Congress—adversaries in 1960—and now look what happened—Whoever thought such a hideous thing could happen in this country—I know how you must feel—so closely missing the greatest prize—and now for you . . . the question comes up again—and you must commit all you and your family's hopes and efforts again—Just one thing I would say to you—if it does not work out as you have hoped for so long—please be consoled by what you already have—your life and your family—We never value life enough when we have it—and I would not have had Jack live his life any other way—though I know his death could have been prevented and I will never cease to torture myself with that—But if you do not win—please think of all that you have."

★

She confided to a friend that she suffered loneliness and despair:

"I'm a living wound. My life is over. I'm dried up—I have nothing more to give, and some days I can't even get out of bed. I cry all day and all night until I'm so exhausted I can't function. Then I drink."

★

I will never go there [Texas] again. You can't imagine how I felt when I was going through Jack's things in the White House and found a set of cuff links in his drawer, emblazoned with the map of Texas. Oh, God—it's awful. I try not to be bitter, but I know I am."

⋆

*J*ack was the love of my life. No one will ever know a big part of me died with him."

⋆

*W*hen she sold Wexford, their house in the Virginia hunt country, she swore the new owners to secrecy, asking them to sign a legal contract prohibiting them from allowing any publicity:

"It's the only house Jack and I ever built together, and I designed it all myself. I don't want it to be exploited and photographed all over the place just because it was ours."

⋆

Mrs. Kennedy decided to keep the house at Hyannis Port and gave the following account of her reasons to Rose Kennedy for her book, *Times to Remember*, published in 1974:

> "We didn't have our own house here until Caroline was born, or what I mean to say is that's when we moved into it. Grandpa wanted to keep everyone together here. I fought against the idea, I thought it was too close, I wanted to be away from the compound.
>
> "But now I am glad. I was reading about a Harvard study of what makes for happy families. Especially what would count most in this age of uncertainty. There were many factors, of course, but close to the top would be a situation in which a number of families knew each other well and had ideas and values they shared, and the children could play at one another's house and sometimes be invited and welcomed for meals. That could happen in a village or small town or a neighborhood, and it's kind of what happens here with the cousins. They're separated more or less most of the year, they live in different cities or different areas of big cities, but they all know one another;

and this is the place where they get together, and
I think that's awfully important for them. . . .

C a m e l o t

Shortly after President Kennedy was buried, she sat down with author Theodore White for an article to be published in *Life* magazine:

"When Jack quoted something, it was usually classical, but all I could keep thinking of is this line from a musical comedy. At night, before we'd go to sleep, Jack liked to play some records. The lines he loved to hear were . . ." and she quoted the lyrics from *Camelot:*

> *Don't let it be forgot*
> *That once there was a spot*
> *For one brief shining moment*
> *That was known as Camelot.*

Then she concluded:

"And it will never be that way again. There'll be great presidents again, but there'll never be another Camelot."

★

*I*n another interview with Theodore White in December 1963: "Once, the more I read of history the more bitter I got. For a while I thought history was something that bitter old men wrote. But then I realized history made Jack what he was. You must think of him as this little boy, sick so much of the time, reading in bed, reading history, reading the Knights of the Round Table, reading Marlborough. For Jack, history was full of heroes. And if it made him this way—if it made him see the heroes—maybe other little boys will see. Men are such a combination of good and bad. Jack had this hero idea of history, this idealistic view."

Future Plans

*B*efore President Kennedy's assassination, she had been asked what she would do after leaving the White House:

"I'll just retire to Boston and try to convince John Jr. that his father was once the president."

★

After the assassination, she planned to live in Washington:

"I'm never going to live in Europe. I'm not going to travel extensively abroad. That's a desecration. I'm going to live in the places I lived with Jack. In Georgetown and with the Kennedys at the Cape. They're my family. I'm going to bring up my children. I want John to grow up to be a good boy."

★

I'm not going to be the Widow Kennedy—and make some speeches like some people who talk about their family. When this is over I'm going to crawl into the deepest retirement. . . . I'm going to live in the place I lived with Jack. . . . That was the first thing I thought that night—where will I go? I wanted my old house back. . . . Then I thought— how can I go back there to that bedroom?"

★

There's only one thing I can do in life now—save my children. They've got to grow up without thinking back at their father's murder. They've got to grow up intelligently, attuned to life in a very important way. And that's the way I want to live my life too."

★

They will never drag me out like a little old widow like they did Mrs. Wilson when President Wilson died. I will never be used that way. . . . I don't want to go out on a Kennedy driveway to a Kennedy airport to visit a Kennedy school. . . . I'm not going around accepting plaques. I don't want medals for Jack. I don't want to be seen by crowds. The first time I minded the crowd was when I went out with the Irish Mafia to the grave."

★

I don't want to be ambassador to France or Mexico. President Johnson said I could have anything I wanted. I would like to work for somebody. . . . I left Washington because of the old haunts. I just couldn't bear to be reminded all the time. I wanted to move into the house I had lived in when Jack was a senator, but I could not get it because someone else had it. . . .

"I offered Jack peace, tranquillity, and serenity, but now the board has moved, all the little pieces changed places. . . . People all over ask me to write . . . there are a lot of requests from magazines, which I've barely looked at. . . . They all want me to write about gracious living or fashion—but I am interested in the same things Jack was interested in."

William Manchester's Death of a President

In early 1964, Jackie and the Kennedy family had asked author William Manchester, a great admirer of President Kennedy, to write the story of the assassination. Manchester agreed to give Jacqueline and Bobby Kennedy editorial control over the completed manuscript in exchange for cooperation. A large percentage of the royalties would be donated to the JFK Library.

However, quite a bit of controversy ensued

over this book. Jackie spoke of it in her oral history for the Lyndon Baines Johnson Library:

". . . in a shell of grief . . . and it's rather hard to stop when the floodgates open. I just talked about the private things. Then the man [Manchester] went away, and I think he was very upset during the writing of the book. I know that afterwards there were so many things . . . which were mostly expressions of grief of mine and Caroline's that I wanted to take out of the book. And whether or not they got out, they were all printed around. Now it doesn't seem to matter so much, but then I had such a feeling.

"I know that everybody else wanted the political things unfair to President Johnson out. And the way that book was done—now, in hindsight—it seems wrong to have ever done that book at that time. Don't forget, these were people in shock. Before we moved out of the White House, Jim Bishop was saying he was going to write a book. . . . All these people were going to do these things, and you thought maybe to just not have this coming up, getting more and more sensational. Choose one person, ask everybody to just speak to him, maybe that would be the right thing to do. Well, it turned out not to be."

There were many aspects of the book that Mrs. Kennedy and Manchester fought over. She had asked for numerous changes and deletions, and was particularly upset about the serialization in *Look* magazine:

"When my children grow up, I don't want them to read all the gruesome stuff about his brain and the way he looked."

Commenting further on the Manchester interviews, she said she spoke freely and in detail because she assumed she was talking primarily "for some scholar in the year 2000."

She did not believe at the time of the interview that any of her statements would be used in the book without her express permission, adding:

"I thought that it would be bound in black and put away on dark library shelves."

On November 28, 1966, Mrs. Kennedy wrote to Manchester in London:

"The changes I am talking about . . . all touch upon things of a personal nature that I cannot bear to be made public. There are many other matters, I know, but these are all of that sort, and they are absolutely necessary to me and my children. I cannot believe that you will not do this much."

*

In January 1967 the case was finally settled out of court when Manchester offered to turn an even larger share of his earnings from the book over to the Kennedy Library.

In 1968 the publisher, Harper & Row, sent a $750,000 check to the library representing the first year's earnings. Mrs. Kennedy issued a statement to the *New York Times:*

"I think it is so beautiful what Mr. Manchester did . . . all the pain of the book and now this noble gesture, of such generosity, makes the circle come around and close with healing."

Memorials

The day before she left the White House, Jackie ordered an inscribed bronze plaque to be placed over the fireplace in the president's bedroom, just beneath a plaque commemorating Abraham Lincoln's occupancy of the same room:

"In this room lived John F. Kennedy with his wife Jacqueline during the two years, ten months and two days he was President of the United States."

The plaque was later removed.

On March 17, 1964, Mrs. Kennedy mailed nine hundred thousand black-bordered prayer cards to acknowledge the sympathy messages she had received:

"I felt St. Patrick's Day was the appropriate time to acknowledge those letters."

She wrote an essay, titled "A Memoir," published on November 27, 1964, in *Look* magazine, a year after President Kennedy's death:

"It is nearly a year since he has been gone.

"On so many days—his birthday, an anniversary, watching his children running to sea—I have thought, 'But this day last year was his last to see that.' He was so full of love and life on all those days. He seems so vulnerable now, when you think that each one was a last time.

"Soon the final day will come around again—as inexorably as it did last year. But expected this time.

"It will find some of us different people than we were a year ago. Learning to accept what was unthinkable when he was alive, changes you.

"I don't think there is any consolation. What was lost cannot be replaced.

"Someone who loved President Kennedy, but who had never known him, wrote to me this winter: 'The hero comes when he is needed. When our belief gets pale and weak, there comes a man out of that need who is shining—and everyone living reflects a little of that light—and stores some up against the time when he is gone.'

"Now I think that I should have known that he

was magic all along. I did know it—but I should have guessed it could not last. I should have known that it was asking too much to dream that I might have grown old with him and see our children grow up together.

"So now he is a legend when he would have preferred to be a man. I must believe that he does not share our suffering now. I think for him—at least he will never know whatever sadness might have lain ahead.

"He knew such a share of it in his life that it always made you so happy whenever you saw him enjoying himself. But now he will never know more—not age, nor stagnation, nor despair, nor crippling illness, nor loss of any more people he loved. His high noon kept all the freshness of the morning—and he died then, never knowing disillusionment.

"He is free and we must live. Those who love him most know that 'the death you have dealt is more than the death which has swallowed you.'"

★

Whenever you drive over the bridge from Washington to Virginia, you see the Lee mansion on the side of the hill in the distance. When Caroline was very little, the mansion was one of the first things she learned to recognize. Now at night you can see [Jack's] flame beneath the mansion from miles away."

★

Mrs. Kennedy was invited to the White House in 1971 by Mrs. Nixon for the ceremonial presentation of the official portraits of John and Jacqueline Kennedy:

"As you know, the thought of returning to the White House is difficult for me. I really do not have the courage to go through an official ceremony, and bring the children back to the only home they both knew with their father under such traumatic conditions. With the press and everything, things I try to avoid in their lives, I know the experience would be hard on them and not leave them with the memories of the White House I would like them to have."

Consequently Mrs. Kennedy and her children came to the White House only for a private viewing.

★

Afterward, Mrs. Kennedy wrote to President and Mrs. Nixon:

"You were so kind to us yesterday. Never have I seen such magnanimity and such tenderness.

"Can you imagine the gift you gave me to return to the White House privately with my little ones while they are still young enough to rediscover their childhood—with you both as guides—and with your daughters, such extraordinary young women.

"What tribute to have brought them up like that in the limelight. I pray I can do half the same with my Caroline. It was good to see her exposed to their example, and John to their charm!

"Thank you with all my heart. A day I always dreaded turned out to be one of the most precious ones I have spent with my children."

Memories of Jack

Mrs. Kennedy moved back to Georgetown after the assassination. She wanted to maintain continuity for the children, and she asked her husband's special assistant, David Powers, to play soldier with John, as he used to do in the White House:

"He'll remember his father through associations with people who knew Jack well."

★

After the assassination, busloads of tourists arrived at her home in Georgetown:

"They actually sit there and eat their lunch and throw sandwich wrappers on the ground. I'm trapped in that house and can't get out. I can't even change my clothes in private because they can look in my bedroom window."

★

One must not let oneself be overwhelmed by sadness."

★

To Joan Braden, wife of newspaper publisher Tom Braden: "There'll never be another Jack. I now understand why he lived so intensely and on the brink. And I'm glad he did."

★

After leaving the White House, she met with decorator Billy Baldwin to discuss decorating her new house in Georgetown. She showed him pieces of Greek sculpture and Roman fragments:

"I have some beautiful things to show you. These are the beginnings of a collection Jack started. It's so sad to be doing this. Like a young married couple fixing up their first house together. I could never make the White House personal. . . ."

She became tearful. "Oh, Mr. Baldwin, I'm afraid I'm going to embarrass you. I just can't hold it in any longer." She buried her face in her hands and wept. Later she continued: "I know from my very brief acquaintance with you that you are a sympathetic man. Do you mind if I tell you something? I know my husband was devoted to me. I know he was proud of me. It took a very long time for us to work everything out, but we did, and we were about to have a real life together. I was going to campaign with him. I know I held a very special place for him—a unique place. . . ." She talked on about JFK and their life. "Can anyone understand how it is to have lived in the White House and then, suddenly, to be living alone as the president's widow? There is something so final about it. And the children. The world is pouring terrible adoration at the feet of my children and I fear for them, for this awful exposure. How can I bring them up normally? We would never even have named John after his father had we known. . . ."

★

About that same time, she sent a photograph of herself with the president to Mrs. J. D. Tippitt, widow of the police officer shot by Oswald in Dallas, and inscribed it:

"There is another bond we share. We must remind our children all the time what brave men their fathers were."

★

In 1964, to a photographer who had taken happy family pictures at Hyannis Port: "Remember then we said the pictures would be so wonderful for the children to look at twenty years from now. But who could know?"

Also in 1964 she said: "I try not to be bitter. I never had or wanted a life of my own. Everything centered around Jack. I can't believe that I'll never see him again. Sometimes I wake in the morning, eager to tell him something, and he's not there. . . . Nearly every religion teaches there's an afterlife, and I cling to that hope. Those three years we spent in the White House were really the happiest time for us, the closest, and now it's all gone. Now there is nothing, nothing."

*

She spoke to her personal secretary, Mary Gallagher, in early December 1963:

"Why did Jack have to die so young? Even when you're sixty, you like to know your husband is there. It's so hard for the children. Please, Mary, don't ever leave. Get yourself fixed for salary on my government appropriation—just don't leave me!"

(She later dismissed her secretary when she moved to New York.)

✳

The world has no right to Jack's private life with me. I shared all these rooms with him, not with the Book-of-the-Month Club readers, and I don't want them shopping through those rooms now."

✳

He lived at such a pace because he wished to know it all."

✳

She spoke with *New York Post* publisher Dorothy Schiff during Robert Kennedy's senatorial campaign in 1964:

"People tell me that time will heal. How much time? Last week I forgot to cancel the newspapers and picked them up and there was the publication of the Warren Report, so I canceled them for the rest of the week. But I went to the hairdresser and picked up *Life* magazine and it was terrible. There is November to be gotten through. . . . Maybe by the first of the year. . . ."

Summer 1964: "I was so fiercely loyal to him I once said thoughtlessly he should be president for life! No, he said, even if he were reelected, eight years in the Presidency is enough for any man."

In March 1964 she showed Richard B. Stolley of *Life* a photo taken from behind of her, President Kennedy, and the children:

"That was only nine days before Jack was killed. It was the only picture I could have around for months. I just couldn't look at his face."

To Dorothy Schiff of the *New York Post,* she reminisced: "I never told him anything or showed him anything unpleasant, and when he got home, I always had his favorite drink, a daiquiri, ready for him, and his favorite record playing, and perhaps a few friends."

★

Jackie decided not to vote in the 1964 election in remembrance of her husband:

"I know that [LBJ] was hurt that I didn't vote in 1964. People in my own family told me I should vote. I said, 'I'm not going to vote.' This is very emotional, but . . . I'd never voted until I was married to Jack. I guess my first vote was probably for him for senator. . . . Then this vote would have been—he would have been alive for that vote. And I thought, 'I'm not going to vote for any[one] because this vote would have been his.'"

★

Her hairdresser at Kenneth's remembers the first anniversary of John F. Kennedy's assassination.

Down Fifth Avenue there were pictures of him in every store window. Mrs. Kennedy arrived, broke down, and sobbed:

"Oh, Rosemary, it was so awful in Washington. They'd follow me everywhere and sit out there in front of the house all day and eat their lunch and throw papers on the lawn. I thought that moving to New York would make it easier for me. If God had only let my baby live. I walk down the street and see his picture in every window. I can't stand it. Why do they remember the assassination? Why can't they celebrate his birthday?"

She told a relative that no matter what she did, she felt overshadowed by death:

"I can't escape it. Whether I'm helping with the Kennedy Memorial at Harvard or taking a plane from Kennedy Airport or seeing a Kennedy in-law, I always think of Jack and what they did to him."

The John F. Kennedy Library

An exhibit of President Kennedy's mementos was to tour the country to raise funds for the Kennedy Library to be built at Harvard:

"I have added some books which he always kept in the Oval Room in the White House. He had them when we were married, and they too give an insight to what he really loved.

"There are books which he read and reread on some of the American statesmen, such as Clay, Calhoun, Webster, John Randolph, and John Quincy Adams. I hope that young people will be interested in seeing these books which the president loved, and that like him they will read American history."

*

She spoke the above words in January 1964. In May of that year she announced the start of the touring exhibit for the library, which would open in Boston in 1979:

"I want to take this opportunity to express my appreciation for the hundreds of thousands of messages . . . which my children and I have received over the past few weeks.

"The knowledge of the affection in which my husband was held by all of you has sustained me, and the warmth of these tributes is something I shall never forget. Whenever I can bear to, I read them. All his bright light gone from the world. All of you who have written to me know how much we all loved him and that he returned that love in full measure.

"It is my greatest wish that all of these letters be acknowledged. They will be, but it will take a long time to do so. . . . I know you will understand.

"Each and every message is to be treasured, not only for my children, but so that future generations will know how much our country and people in other nations thought of him. Your letters will be placed with his papers in the library to be erected in his memory. I hope that in years to come many of you and your children will be able to visit the Kennedy Library. It will be, we hope, not only a memorial . . . but a living center of study . . . for young people and for scholars from all over the world.

"May I thank you again on behalf of my children and of the president's family for the comfort that your letters have brought to us all."

On the library: "I wanted people to see the rather personal side of the president, so I have parted with some of our greatest treasures, the pictures, objects, and books, which we always kept at home . . . [to] give an insight to what he really loved."

We want this library to be . . . a place which can keep alive all the things he stood for. . . . Many people in many countries have written to me, saying that he gave them new confidence in America, and in their own ability to solve their own problems. . . . The deep desire to inspire people, to take an active part in the life of their country . . . attracts our best people to political life. . . . We should all do something to right the wrongs we see and not just complain about them. We owe that to our country. . . . [It] will suffer if we don't serve it. . . . That's why we're working to build the library. . . . His office will be there. . . . You can hear every speech he made, and movies. You can see all the manuscripts of his speeches and how he changed them."

★

Mrs. Kennedy made a film for the exhibit to raise money for the JFK Library. She was nervous appearing before the camera:

"I wish I knew when to breathe. I just don't see how actresses can do this."

The Assassination of Martin Luther King Jr.

In April 1968, just four years after President Kennedy's death, the nation was rocked once more by the assassination of Martin Luther King Jr. Mrs. Kennedy wrote to Coretta Scott King:

"Your husband was one of the greatest and most inspiring leaders that any of us has known. I share your grief in this hour of sorrow but his death will help to free us from the violence and tragedy which hate often produces. He will always be remembered as one of our nation's martyred greats."

The Assassination of Robert F. Kennedy

By that time, Robert Kennedy had decided to run for president. Jackie had grave concerns about his safety and expressed those concerns on more than one occasion. At a New York dinner party that month, Jackie took the historian Arthur Schlesinger aside and said:

"Do you know what I think will happen to Bobby? The same thing that happened to Jack. . . . There is so much hatred in this country, and more people hate Bobby than hated Jack. . . . I've told Bobby this, but he isn't fatalistic, like me."

★

After Bobby's death, she went to the hospital in Los Angeles and spoke to his press secretary, Frank Mankiewicz:

"The Church is . . . at its best only at the time

of death. The rest of the time it's often rather silly little men running around in their black suits. But the Catholic Church understands death. I'll tell you who else understands death are the black churches. I remember at the funeral of Martin Luther King I was looking at those faces, and I realized that they know death. They see it all the time and they're ready for it . . . in the way in which a good Catholic is. We know death. . . . As a matter of fact, if it weren't for the children, we'd welcome it."

After Robert Kennedy's assassination she told a friend: "I despise America and I don't want my children to live here anymore. If they're killing Kennedys, my kids are number one targets. . . . I want to get out of this country."

CHILDREN
AND FAMILY

Raising children is the best thing I've ever done. Being a mother is what I think has made me the person I am."

★

Most men don't care about children as much as women do, but [Jack] did. He was the kind of man who should have had a brood of children."

★

[Jack] always wanted a baby coming along when its predecessor was growing up. That is why he was so glad when he learned that I was having Patrick. But he wished for five children. Before we were married he said that. And he had four children in seven years."

All the places and feelings and happiness that bind you to a family you love are something that you take with you no matter how far you go."

In 1960 she was worried about the effect on the children of their father's nomination and subsequent campaign:

"I worry. All those books on child psychology— and I'm the type who reads all those books—talk about how things affect children Caroline's age. I get this terrible feeling that when we leave, she might think that it's because we don't want to be with her. After the convention, Jack was here for three straight weeks, and Caroline got so used to having Daddy around the house."

★

Toronto Daily Star, November 9, 1960: "A big problem in my life . . . has been that campaigning with Jack has taken me away from home. If Jack proved to be the greatest president of the century and his children turned out badly, it would be a tragedy."

★

She said upon moving into the White House: "I want my children to be brought up in more personal surroundings, not in the staterooms. And I don't want them to be raised by nurses and Secret Service agents."

★

It isn't fair to children in the limelight to leave them to the care of others and expect they will turn out all right."

★

I think it's hard enough to bring up children anyway, and everyone knows that limelight is the worst thing for them. They either get conceited, or else they get hurt. . . . They need their mother's affection and guidance and long periods of time alone with her. That is what gives them security in an often confusing new world."

★

People have too many theories about rearing children. I believe simply in love, security, and discipline."

★

I always imagined I'd raise my children completely on my own. But once you have them, you find you need help. So I need Dr. Spock a lot, and I find it such a relief to know that other people's children are as bad as yours at the same age."

★

On developing creativity in children: "Perhaps some painting—just some splattering of watercolors or crayon lines the way a child loves to do it—is the first step. Whenever I paint now, I put up a child's paint box for Caroline beside me. She really prefers to dip the brushes in water, smear the paints, and make a mess, but it is a treat for her to paint with her mother.

"Perhaps this will develop a latent talent; perhaps it will merely do what it did for me, produce occasional paintings which only one's family could admire, and be a source of pleasure and relaxation."

✷

I hope I do as well for my children as my mother did for me."

✷

*A*s long as the father is the figure of authority, and the mother provides love and guidance, children have a pretty good chance of turning out all right. The family is the prime unit in society. Unless its ties are loosened, children can be properly reared."

✷

The things you do with your children, you never forget."

★

There are many little ways to enlarge [your child's] world. Love of books is the best of all."

★

Most important, of course, is not to shut off the inquiring mind by being impatient with its questions. The seemingly endless chain of 'whys' means something important to the child—his way of learning."

★

After President Kennedy was killed, danger was never far from Jackie's mind. She once confided to a teacher:

"I'm nerve-racked about the safety of the children. There are so many nutcases about."

The first family celebrates Caroline's birthday. (Photo by Cecil W. Stoughton. Courtesy of the John Fitzgerald Kennedy Library, Boston, Massachusetts.)

★

*I*f you bungle raising your children, I don't think whatever else you do well matters very much. And why shouldn't that be an example too?"

★

*T*o Hillary Clinton during a boat ride around Martha's Vineyard in the summer of 1993—her rule for raising Caroline and John:

"Don't let them have too much attention or be exposed too much, because they deserve a chance to grow up to be who they're going to be."

Caroline Bouvier Kennedy

*O*n November 29, 1957, the day after Thanksgiving, Caroline Bouvier Kennedy was born at New York's Lying-In Hospital, Cornell Medical Center. Three weeks later, wrapped in the same robe her mother had worn at her own christening, Caroline was baptized by Cardinal Cushing at St. Patrick's Cathedral.

★

She was born four months after the death of her grandfather "Black Jack" Bouvier. Jacqueline regretted that her father didn't live to see his first grandchild:

"He would have been so happy, so happy. [I made Jack promise me] that whether [it was] a boy or a girl, we [would] give the baby the name of Bouvier."

★

[Jack] is so affectionate with his daughter. She has made him so much happier. A man without a child is incomplete."

★

During the campaign, Caroline's first spoken words were "good-bye," "New Hampshire," "Wisconsin," and "West Virginia."

Jackie said:

"I am sorry so few states have primaries, or we

would have a daughter with the greatest vocabulary of any two-year-old in the country."

*F*rom a thank-you note during the White House years to a photographer who agreed not to publish photos of Caroline: "I know that newspapers need to print different—or rather unusual—pictures and there is the conflict of trying to raise one's children fairly normally. So when you—who are torn both ways—respect a little girl's chance to have a happy day with the other children who fortunately treat her as just another 4-year-old (that is almost the only public place where she isn't singled out and fawned over)—it's amazing and consoling."

*T*o a close friend and fellow parent in 1970 in New York: "Do you have the same problems with your girls as I have with Caroline? She knows everything and I don't know anything. I can't do anything with her."

★

Quoted by the designer Carolina Herrera, referring to Caroline's wedding dress design: "I am not
going to get involved because Caroline is the one
who will wear it. I want her to be the happiest girl in
the world."

John F. Kennedy Jr.

John F. Kennedy Jr. was born on November 25,
1960, just after Thanksgiving, at Georgetown
University Hospital, in Washington, D.C. His father
had just been elected president.

★

Just after John F. Kennedy Jr.'s birth, reporters
asked if she'd like more children:
"I'd be delighted. I hope that I have many
more."

★

I don't want the children to be just two kids living on Fifth Avenue and going to nice schools. There's so much else in the world, outside this sanctuary we live in. Bobby has told them about some of those things—the children of Harlem, for instance. He told them about the rats and about the terrible living conditions that exist right here in the midst of a rich city. Broken windows letting in the cold. John was so touched by that he said he'd go to work and use the money he made to put windows in those houses. The children rounded up their best toys last Christmas and gave them away.

"I want them to know about how the rest of the world lives, but also I want to be able to give them some kind of sanctuary when they need it, some place to take them into when things happen to them that do not necessarily happen to other children."

★

Caroline is more withdrawn, but John, well, he's something else. John makes friends with everyone. Immediately. He surprises me in so many ways. He seems so much more than one would expect of a child of six. Sometimes it almost seems that he is trying to protect me instead of just the other way around.

"There was that day last November, the day of the anniversary. As the two of us walked home from school, I noticed that a little group of children, some of them from John's class, was following us. Then one of the children said, quite loud, 'Your father's dead . . . your father's dead!' You know how children are. They've even said it to me when I've run into them at school, as if . . . Well, this day John listened to them saying it over and over, and he didn't say a word. He just came close to me, took my hand, and he squeezed it. As if he were trying to reassure me that things were all right. And so we walked home together, with the children following us.

"I sometimes used to say to myself, 'He'll never remember his father. He was too young.' But now I think he will. He'll remember his father through associations with people who knew Jack well and the things Jack liked to do. He will be getting to know his father. I tell him little things like, 'Oh, don't worry about your spelling, your father couldn't spell very well either.' That pleases him, you can bet.

"I want to help him go back and find his father.

It can be done. There was that stone his father placed on a mound during his visit to Argentina a long time ago, and then when I took the children there later, John put a stone on top of his father's. He'd like to go back to Argentina and see his stone, and his father's stone—and that will be part of knowing his father."

Patrick Bouvier Kennedy

Patrick Bouvier Kennedy was born five weeks early on August 7, 1963, at Otis Air Force Base, Cape Cod. Suffering from hyaline membrane disease, he was transferred to Children's Medical Center in Boston. He died after only three days of life.

✶

After the death of her second son: "Oh, Jack, there's only one thing I could not bear now—if I ever lost you."

✶

Jackie said later that the president had wanted another boy:

"He felt the loss of the baby in the house as much as I did."

[Patrick's death] was the only time I ever saw him [JFK] cry. He was inconsolable. As shocking as it was for me, it was worse for him. Jack nearly collapsed over it. Although he never said so, I know he wanted another boy—John was his real kin spirit."

Leaving the hospital afterward, she thanked the nurses:

"You've been so wonderful to me that I'm coming back here next year to have another baby. So you better be ready for me."

MARRIAGE TO ARISTOTLE ONASSIS

After the loss of baby Patrick in 1963, Jackie's sister, Lee, confided in longtime friend Aristotle Onassis. Ari suggested that Jackie and Lee visit Greece to cruise the islands. He offered to put his yacht, the *Christina,* and its crew at their disposal. They could bring friends and travel wherever they wished. He offered to remain ashore or out of sight, whichever they preferred. Mrs. Kennedy accepted the invitation but insisted that Ari accompany them:

"I can't possibly accept this man's hospitality and then not let him come along. It would be too cruel."

★

In October 1968, Mrs. Kennedy married Onassis. "Jackie, you're going to fall off your pedestal," friends warned her.

"That's better than freezing there," she reportedly replied.

★

Regarding her remarriage, she told Truman Capote: "I can't very well marry a dentist from New Jersey."

★

At the time of her wedding to Onassis, the press landed on the beach at Skorpios. Onassis's seamen met them, and it became obvious that someone might get hurt. Concerned about everyone's well-being, the bride-to-be called off Onassis's force, saying about the media: "These men also have to make a living."

★

On the morning of the wedding, October 20, 1968, an attempt was made to make peace with the media. She wrote to the press:

"We know you understand that even though people may be well known, they still hold in their hearts the emotions of a simple person for the moments that are the most important of those we know on earth—birth, marriage, and death. We wish our wedding to be a private moment among the cypresses of Skorpios with only members of the family present—five of them little children. If you will give us these moments, we will gladly give you all the cooperation possible for you to take the pictures you need."

She would later comment that marrying Ari helped restore peace in her life:

"Nobody could understand why I married Ari. But I just couldn't live anymore as the Kennedy widow. It was a release, freedom from the oppressive obsession the world had with me."

✫

Marrying him [Onassis] liberated me from the Kennedys especially the Kennedy administration. None of them could understand why I'd want that funny, little squiggly name when I used to have the greatest name of all."

✫

Jackie was the target of a certain amount of hate mail throughout her public life. During her marriage to Onassis, she had round-the-clock protection:

"I guess the theme song of my life is that oldie 'Me and My Shadow.'"

✫

Shortly after Robert Kennedy's assassination, Jackie said of her marriage to Onassis:

"I wanted to go away. They were killing Kennedys and I didn't want them to harm my children. I wanted to go off. I wanted to be somewhere safe."

*Jackie and Ari enjoy the scenery from aboard a boat during a ten-day
tour of Egypt. (© Bettmann/CORBIS)*

✳

About her husband's work ethic, Jackie quipped, "Ari never stops working. He dreams in millions."

✳

When Truman Capote once told Jackie, "I gave a party and my dog chewed up Lee Radziwill's sable coat," Jackie responded: "Don't worry. We can buy another sable for Lee tomorrow and charge it to Ari. He won't mind."

✳

When Ari saved Jackie's cousin Edie Beale and her mother from eviction by paying for the repairs on their Long Island house, Jackie was quoted as having said, "Don't you think I'm lucky to be married to such a splendid person?"

✳

For Jackie's fortieth birthday, Onassis gave her, among other gems, a pair of "Apollo 11" earrings to celebrate Neil Armstrong's walk on the moon. At the party, Greek actress Katina Paxinou sat alongside Mrs. Onassis and complimented her on her new earrings. Two spheres representing earth and the moon were joined by what was supposed to be a miniature spaceship:

"Ari was actually apologetic about them. He felt they were such trifles. But he promised me that, if I'm good, next year he'll give me the moon itself."

*R*egarding Onassis's taste in clothes:

"Look at him. He must have four hundred suits. But he wears the same gray one in New York, the same blue one in Paris, and the same brown one in London."

*A*s for the many rumors that spread after her marriage to Onassis, Jackie told Truman Capote: "It's a lie, a complete lie. I don't have any money. When I married Ari, my income from the Kennedy estate stopped and so did my widow's pension from the

U.S. government. I didn't make any premarital financial agreement with Ari. I know it's an old Greek custom, but I couldn't do it. I didn't want to barter myself. Except for my personal possessions, I have exactly five thousand two hundred dollars in a bank account. Everything else I charge to Olympic Airways."

★

In 1975, when Ari was very ill, Mrs. Onassis flew to Athens with the heart specialist Dr. Isadore Rosenfeld. Jackie was firm about having her husband moved to Paris for his care.

"He's my husband and I believe this switch is necessary. Let's not argue about it."

★

In Athens after Onassis's funeral, she gave a short statement to the press:

"Aristotle Onassis rescued me at a time when my life was engulfed in shadows. He meant a lot to me. He brought me into a world where one could find both happiness and love. We lived through many beautiful experiences together which cannot be forgotten, and for which I will be eternally grateful. . . . Nothing has changed both with Aristotle's sisters and his daughter. The same love binds us as when he lived."

She returned to Greece to dedicate a new wing of a children's camp in memory of Onassis:

"My main purpose for coming to Greece, apart from loving the country, is to put into practice the last instructions of my late husband in order to preserve his name."

ON HER
OWN AGAIN

Having been widowed twice by the age of forty-five, Jackie confided to a friend: "I have always lived through men. Now I realize I can't do that anymore."

⋆

After Ari's death, she had a number of prominent escorts, one of whom was Mike Nichols.

"Taking you anyplace is like going out with a national monument," he once said.

"Yes," Mrs. Onassis retorted, "but isn't it fun?"

⋆

As she was leaving a dinner party in New York one evening, Mr. and Mrs. Bertrand Taylor, who lived just a few blocks north of Mrs. Onassis on Fifth Avenue, offered her a lift home. She accepted, and when Mr. Taylor went outside to get the car, he noticed two men with cameras standing in the shadows. When he went back inside, he said, "I think I'd better warn you that there are two photographers outside."

With a look of mock anguish, she replied, "Only two? I must be slipping!" With a wink she added, "Let's have some fun with them."

Smiling, she put her hand in the crook of Taylor's elbow and walked onto the sidewalk with him.

★

Paris-Match correspondent Benno Graziani knew Federico Fellini well. Once when Fellini was in New York, he asked if Benno could arrange for him to meet Mrs. Onassis. When they were dining together, she spoke in detail of Fellini's work. Benno commented to her, "I didn't know you were so familiar with Fellini's work."

She replied, "When I knew he was coming for dinner, I watched all his films and read practically all the books written about him."

∗

While living in New York during the early 1970s, Mrs. Onassis would answer the phone on her maid's day off with a fake Spanish accent in the hope that callers wouldn't recognize her voice:

"I have to do that to get rid of people."

∗

Jackie also liked to people-watch:

"Think of the plots that are being hatched down there," she said once as she looked down from the balcony of the Four Seasons restaurant.

∗

Over lunch, André Previn asked her whether it bothered her that people looked at her:

"That's why I always wear my dark glasses. It may be that they're looking at me, but none of them can ever tell which ones I'm looking back at. That way I can have fun with it!"

★

During her years in New York, Jackie became involved in the fight to save Grand Central Station:

"We've heard that it's too late to save this station, but that's not true. Even at the eleventh hour you can succeed, and I think that's exactly what we'll do. I care desperately about saving old buildings."

★

She also cared deeply about preserving what remained of nature in city life. When she first heard about the construction of the Columbus Center Building in Manhattan, which was to be the tallest building in the world, she lamented to Kent L. Barwick, president of the Municipal Art Society:

"They're stealing the sky."

Work

In 1975, when her old friend Letitia Baldrige suggested that Jackie get a job, she replied: "Who me—*work?*"

*

But soon after, she joined Viking Publishers as a consulting editor, moving on to Doubleday in 1977:

"After I got out of college, I wanted to write for a newspaper or work for a publishing house, but I did other things. When the time was right, I did this. I would always have liked to. I see my future as staying on as an editor at Viking, hopefully. I love the work I do."

*

In an interview with *Ms.* magazine in March 1979, she discussed her work:

"Before I was married, I worked on a newspaper. Being a journalist seemed the ideal way of both having a job and experiencing the world, especially for anyone with a sense of adventure. I wouldn't choose it as a profession now—journalism has variety but doesn't allow you to enter different worlds in depth, as book publishing does—though I understand why so many young people are attracted to it. Being a reporter seems a ticket out to the world.

"If I hadn't married, I might have had a life very much like Gloria Emerson's. She is a friend who started out in Paris writing about fashion—and then ended up as a correspondent in Vietnam. The two ends of her career couldn't seem farther apart, and that is the virtue of journalism. You never know where it's going to take you, but it can be a noble life—she became a correspondent and an author of influence.

"What has been sad for many women of my generation is that they weren't supposed to work if they had families. There they were, with the highest education, and what were they to do when the children were grown—watch the raindrops coming down the windowpane? Leave their fine minds unexercised? Of course women should work if they want to."

★

One of the reasons Jackie ended up in publishing was that she had a lifelong love of reading, well before she began her career. Books for grown-ups were kept in the guest room where she took her afternoon naps as a child. One day she remarked, "Mummy, I liked the story of the lady and the dog."

Her mother discovered that Jackie had been reading a book of short stories by Chekhov with sophisticated plots and elaborate Russian names.

"Did you understand all the words?" her mother asked.

"Yes—except what's a midwife?" asked Jackie.

"Didn't you mind all those long names?"

"No, why should I mind?" her six-year-old daughter replied.

★

She came to love and be most knowledgeable about the eighteenth century. She wrote:

"When you read a lot you come across some things that interest you more than others. So you read a little bit more about those things that interest you. I was fascinated by what I read about the eighteenth-century period and as I dug deeper I became more fascinated. First thing I knew, I wanted to know everything I could about the period."

*H*er last interview was with John F. Baker, the British-born editor of *Publishers Weekly,* a little more than a year before her death:

"I love books. I've known writers all my life.

"I'm drawn to books that are out of our regular experience. Books of other cultures, ancient histories. I'm fascinated by hearing artists talk about their craft. To me, a wonderful book is one that takes me on a journey into something I didn't know before."

*A*ware that she was not being accepted at first as a professional in the world of publishing, she was defensive:

"It's not as if I've never done anything interesting. I've been a reporter myself and I've lived through important parts of American history. I'm not the worst choice for this position."

✴

What I like about being an editor is that it expands your knowledge and heightens your discrimination. Each book takes you down another path. Hopefully, some of them move people and some of them do some good."

✴

A hurricane would really be good background for a novel, the way people all behave differently in the midst of disaster. Some panic, some get nasty, others are brave. We had to live by candlelight. It's hard to read by candlelight. How did Thomas Jefferson get through reading all that stuff?"

✴

Quoted by author Eugene Kennedy, who worked on a book with her: "Like everybody else," she said good-humoredly, "I have to work my way up to an office with a window."

★

Jackie left Viking in 1977 because of Jeffrey Archer's controversial novel *Shall We Tell the President?* depicting Teddy Kennedy as the target of an assassination attempt. She issued a statement:

"Last spring when told of the book, I tried to separate my lives as a Viking employee and a Kennedy relative. But this fall, when it was suggested that I had had something to do with acquiring the book and that I was not distressed by its publication, I felt I had to resign."

★

After leaving Viking, she moved on to Doubleday:

"One of the good things about working for a publishing house like Doubleday is its size: somewhere among its different divisions, there is a place for almost every subject and kind of book."

★

As an editor, Jackie had a line to draw between the need for publicity and her own reserve. Judith

Martin (Miss Manners) covered a book party for *In the Russian Style,* which Jackie had edited for Viking Press. The book consisted of photographs of and quotations about imperial Russian society, including their clothes, jewels, lavish parties, marriages, and love affairs, with judgments on same. The reporter asked the former first lady if her "point of view changed about the propriety of examining the private lives of public figures?"

"These things were public," she replied, referring to the sources, including Catherine the Great's love letters. "When it's past, it becomes history." And if a historian someday uses her letters? "I won't be here to mind."

Mimi Kazon, a former political columnist, met Jackie at a book party and agreed to send her a packet of her best material for possible publication by Doubleday. A few months later she received a call:

"This is Jackie Onassis of Doubleday. I received your columns and found them quick and witty. But they were all about power, and frankly I'm not into power."

*

*A*n editor mentioned during a meeting that he was trying to get a Hunter Thompson book. Sitting next to him, she slipped him a note:

"I would give up food to publish Hunter Thompson."

*

*I*n 1982 she wrote to Louis Auchincloss regarding his book *False Dawn,* a collection of portraits of seventeenth-century women:

"The most uncomfortable thing I have ever had to do is edit your immaculate writing. I hope and expect that you will object vociferously and that I will learn a lot from you in the process.

"Please realize that one gets obsessive and nit picking when editing a manuscript filled with facts, in a concentrated session. I did yours in a day and a night in Martha's Vineyard. It isn't at all like reading a book for pleasure and I may have been overzealous."

*

John Russell, former chief art critic of the *New York Times,* recalls his ambition to write something that would last—forever—as a thank-you letter to the United States. But the idea began to collapse under its own weight. Jackie said to him:

"Don't let's talk anymore about that book you're never going to write."

*

Jackie's writing advice: "Don't allow yourself to be repetitious or sentimental. It backfires."

*

In 1993 she said of working with authors and agents: "I certainly don't think dealing with authors and agents is very hard. I don't work with agents as much as some editors, perhaps—though sometimes, when something crosses their desk, I hope they think of me and say, 'Oh, she might like that.'"

*

*J*ackie clearly found editing very gratifying work: "If you produce one book, you will have done something wonderful in your life."

Art

*M*rs. Onassis once said that the best way to learn to appreciate art is "by using your eyes, by focusing your whole attention on a work of art to try to understand the message the artist wants to convey."

*

[*A*] child of any age gets his own message, his very own important emotional response from looking at a work of art. He should be encouraged to have that opportunity often, by his parents and teachers."

*

*S*howing pictures of Lincoln during the televised White House tour:

"Here is what the White House did to President Lincoln. Here is how he changed: 1861, the strong man with the arched eyebrow; 1865, one week before his assassination."

The four photographs, one for each year of his presidency, show Lincoln's face aging dramatically.

★

Regarding the painting of herself she chose for the White House:

"I would have liked it even more lost in shadows. Less specific, more impressionistic."

★

President Kennedy and I shared the conviction that the artist should be honored by society, and all of this had to do with calling attention to what was finest in America, what should be esteemed and honored. The arts had been treated as a stepchild in the United States. When the government had supported the arts, as in many WPA projects, artists were given a hand, and some wonderful things emerged. . . . Our great museums and great performing companies should of course be supported, but the experimental and the unknown should also be thrown a line. Our contemporary artists—in all the media—have excited the world. It was so sad that we couldn't help them more."

Asked if she enjoyed the ballet—at which the press flashed cameras up until the start of the performance—she replied: "Yes. Once the lights went down."

About a dinner given for French Minister of Culture André Malraux: "[He said] public spaces with sculpture can affect someone rushing off to work, and maybe change their perception of themselves, even slightly."

I had hoped one day to have a minister for the arts in the cabinet. Much groundwork would have to be done before that would be possible."

T r a v e l s

*D*uring Jackie's famous trip to Paris with President Kennedy in June 1961, she told reporters she would have preferred to "walk around and look at the buildings and the streets and sit in the cafés."

*O*n that same trip, Mrs. Kennedy spoke French to Charles de Gaulle, saying, "My grandparents are French."

"So are mine, madame," he replied.

*U*pon her return from France she said, "Let me stay as happy as this forever."

*

After visiting Pope John XXIII in March 1962: "He's such a good man—so of the earth and centuries of kindness in his eyes. I was determined to curtsy three times on the way, as you're supposed to do. I did once, and then he rushed forward so I barely got in one more curtsy. I read in the papers I had an unusually long audience, but it didn't seem long. It was all so simple and natural. We didn't talk of anything serious."

*

On Greece: "This land is a miracle. It's like a dream. . . . I am literally enchanted with your clear blue sky, as well as with your beautiful sea. . . . Everyone should see Greece. . . . My dream is to have a house here to spend vacations with my children."

*

After the trip to Vienna in 1961 for a summit meeting with Khrushchev: "I like [Mrs.

Khrushchev]. She was the kind of woman you'd ask in perfect confidence to baby-sit for you, if you wanted to go out some evening."

★

On meeting Khrushchev: "Oh, Mr. Chairman, don't bore me with statistics."

★

During a goodwill trip to India in 1962 with her sister, Lee, after a sidesaddle ride on an elephant: "A camel makes an elephant feel like a jet plane."

★

On her return to the White House, she said: "Before we left Washington, I said I wanted to spend one night in a village. The date and the village had been decided upon, but somehow it got out of the schedule. You know, the Indians decided our trip, and I don't blame them. They could have shown me their deepest poverty, hoping I'd go home and say they needed more aid, but they were too thoughtful for that. They just wanted us to have magical memories of an enchanted visit."

*

Quoted on leaving Hawaii in July 1967: "I had forgotten and my children have never known what it is like to discover a new place unwatched and unnoticed."

Humor

Jackie had a great sense of humor and often used it to diffuse a situation or lighten things up.

*

In 1953, at a small party on the yacht of Aristotle Onassis, Senator Kennedy played up to Prime Minister Winston Churchill, but Churchill failed to recognize him. As the young couple left, Mrs. Kennedy eyed her husband in his tuxedo and suggested, "Maybe he thought you were the waiter, Jack."

★

When asked the best site for the 1960 Democratic National Convention, Mrs. Kennedy suggested, "Acapulco."

★

During the campaign, a reporter asked, "Is your baby due before inauguration day?"

Jacqueline Kennedy said, "When's inauguration day?"

★

Mrs. Kennedy didn't like her husband to bring his work home. According to a White House butler, the president exclaimed, "What in hell am I ever going to do about air pollution?"

Jackie suggested, "It's very simple, my dear. Get the air force to spray our industrial centers with Chanel No. 5."

★

She opposed the McCarran-Walter Immigration Act, passed in 1952 over the veto of President Truman. The Kennedy administration was working on relaxing its restrictions:

"A bill that restrictive might not have let in the Bouviers or the Kennedys."

＊

Once during the White House years when Mrs. Kennedy turned up with a new German shepherd puppy, the press asked her what she would feed it.

"Reporters," she answered.

＊

Pablo Casals had played in the East Room. He was superb. Months later, still being complimented on the fact that her husband had done so much for music through the Casals performance, Jackie kidded:

"The only music he really appreciates is 'Hail to the Chief'!"

Caroline Kennedy graduates from Radcliff at a joint session during Harvard University's 329th Commencement. John Jr., Senator Edward Kennedy, and Jackie were there to offer congratulations. (© Bettmann/CORBIS)

★

To August Heckscher, special consultant to the president on the arts: "Mr. Heckscher, I will do anything for the arts you want. . . . But, of course, I can't be away too much from the children and I can't be present at too many cultural events. . . . After all, I'm *not* Mrs. Roosevelt."

★

When Harvard men say they have graduated from Radcliffe, then we've made it."

★

Among the photos taken for the White House guidebook was one of the children in John Jr.'s bedroom. Everyone liked the picture, but Mrs. Kennedy would not allow it to be used:

"Gentlemen, even at the age of two one's bedroom should be private."

★

Some would-be kingmakers in New York wanted Jackie to run for the Senate. Her reply:

"If I could do it three days a week."

Jackie on Herself

Although shy and introverted as a young girl, over the years she developed confidence in her own style and viewpoint.

★

On May 25, 1972, she told the *Scranton Times:*

"Why do people always try to see me through the different names I have had at different times? People often forget that I was Jacqueline Bouvier before being Mrs. Kennedy or Mrs. Onassis. Throughout my life I have always tried to remain true to myself. This I will continue to do as long as I live."

★

The trouble with me is I'm an outsider. And that's a very hard thing to be in American life."

✴

I have a tendency to get into a downward spiral of depression or isolation when I'm sad. To go out, to take a walk, to take a swim, that's very much what the Kennedys do. It's a salvation, really."

✴

I'm solitary. I'm rather introverted. I'm really glad my children have a sense of humor—I think I'm a bit irreverent."

✴

I think I'm more of a private person. I really don't like to call attention to anything."

✴

I am happiest when I'm alone."

✶

I am a woman above everything else."

✶

I do love to live in style."

✶

I get afraid of reporters when they come to me in a crowd. I don't like crowds because I don't like impersonal masses. They remind me of swarms of locusts."

✶

*T*he truth of the matter is that I am a very shy person. People take my diffidence for arrogance and my withdrawal from publicity as a sign, supposedly, that I am looking down on the rest of mankind."

✶

I am today what I was yesterday, and with luck, what I will be tomorrow."

*

*A*sked by poet Stephen Spender at a 1979 dinner party to name her proudest accomplishment, she answered: "Well, I think my biggest achievement is that, after going through a rather difficult time, I consider myself comparatively sane."

*

*O*n her place in history: "So many people hit the White House with their Dictaphone running. . . . I never even kept a journal. I thought, 'I want to live my life, not record it.'"

*

*D*oubleday's deputy publisher Bill Barry, who worked with her for many years, suggested that she write her memoirs. She said that life is too precious:

"I want to savor it. I'd rather spend my time

feeling a galloping horse, or the mist of the ocean up at Martha's Vineyard."

★

Author David Wise saw her about two years before her death and asked her when she would write her own book. She laughed and replied:

"Maybe when I'm ninety."

She said that people change and the person she would have written about thirty years ago "is not the same person today. The imagination takes over. When Isak Dinesen wrote *Out of Africa,* she left out how badly her husband had treated her. She created a new past, in effect. And why sit indoors with a yellow pad writing a memoir when you could be outdoors?"

Wise asked her how she managed to deal with the tabloids:

"The river of sludge will go on and on. It isn't about me."

★

When you get written about a lot, you just think of it as a little cartoon that runs along at the bottom of your life—but one that doesn't have much to do with your life."

✴

The only routine with me is no routine at all."

✴

The sensational pieces will continue to appear as long as there is a market for them. One's real life is lived on another private level."

✴

In spite of the fact that there are more article entries under her name in the *Readers' Guide to Periodical Literature* than for any other living American woman, she had never spoken to more than one or two of those reporters and authors.

She told friend Jayne Wrightsman about two years before she died:

"I'm sixty-two now, and I've been in the public

eye for more than thirty years. I can't believe that anybody still cares about me or is still interested in what I do."

[Jack is] the one who warrants—consideration, not me. I could never think otherwise."

Words of Wisdom

In March 1961, she said: "Happiness is not where you think you find it. I'm determined not to worry. So many people poison every day worrying about the next. I've learned a lot from my husband."

Robert Kennedy quoted Jackie after she returned from the funeral of Martin Luther King Jr. in April 1968: "Of course people feel guilty for a moment. But they hate feeling guilty. They can't stand it for very long. Then they turn."

★

If men only knew how great they looked in their white tie and tails, they would wear them every night of their lives."

★

Bruce Tracy, an editor at Doubleday, once had a chance to go to Europe but would have to miss some events regarding the publication of a book they were both working on. He asked Jackie's advice, and she said:

"Life comes first."

★

Joan Kennedy, saddened by the death of a man she might have married after her divorce from Ted Kennedy, asked Jackie, "When does the heartache end? I finally meet a decent man and he's taken from me. It's just not fair."

Jackie replied:

"Joan, do you really expect life to be fair after

everything we've gone through? It's up to you to take what happiness you can find. And you have to soldier on, whether you like it or not."

★

Longtime friend Viviana Crespi remembers that her son adored Mrs. Onassis. He'd sent her his poems, and she replied:

"You must continue. Poets are the ones who change the world."

★

There are two kinds of women, those who want power in the world and those who want power in bed."

★

Regarding New York society women: "These women who seem to have it all . . . are really desperate and trapped."

★

Sex is a bad thing because it rumples the clothes."

★

The first time you marry for love, the second for money, and the third for companionship."

★

If you cut people off from what nourishes them spiritually and historically, something within them dies."

★

If you get out into the world, and move around a bit, you begin to see that there are people who have been through much worse things than I have."

★

I have been through a lot and I have suffered a great deal, as you know. But I have had lots of happy moments as well. I have come to the conclusion that we must not expect too much from life. We must give to life at least as much as we receive from it. Every moment one lives is different from the other, the good, the bad, the hardship, the joy, the tragedy, love and happiness are all woven into one single indescribable whole that is called life. One must not dwell on only the tragedies that life holds for us all, just as a person must not just think of only the happiness and greatness that they've experienced in life. If you separate the happiness and the sadness from each other, then neither is an accurate account of what life is truly like. Life is made up of both the good and the bad—and they cannot be separated from each other."

Maurice Tempelsman

*B*elgian-born financier and diamond merchant Maurice Tempelsman was Jackie's companion during the last years of her life. Dignified, intellectual,

and charming, Tempelsman had first met the
Kennedys in the 1950s. Starting as a friend, he later
became her financial adviser, increasing her fortune
considerably. Along with a shared love of art and
antiques, they spent quiet times sailing on his yacht,
bird-watching, and strolling in Central Park. Jackie
had once said, "I admire Maurice's strength and his
success. I truly hope my notoriety doesn't force him
out of my life."

Maurice did, in fact, remain in her life till the end.
Shortly before her death, she wrote notes to her
children. To John, she wrote:

"I understand the pressure you'll forever have
to endure as a Kennedy, even though we brought you
into this world as an innocent. You, especially, have
a place in history.

"No matter what course in life you choose, all I
can ask is that you and Caroline continue to make
me, the Kennedy family, and yourself proud.

"Stay loyal to those who love you. Especially
Maurice [Tempelsman]. He's a decent man with an

abundance of common sense. You will do well to seek his advice."

Grandchildren

On July 19, 1986, Caroline Kennedy married Edwin Schlossberg. Their first child, daughter Rose, was born June 25, 1988, at New York Hospital, Cornell Medical Center.

★

In the spring of 1988, Mrs. Onassis's common refrain to friends was: "I'm going to be a grandmother—imagine that."

★

About her grandchildren (Rose, then six; Tatiana, three; and Jack, sixteen months): "They make my spirits soar!"

★

To Caroline, she wrote: "The children have been a wonderful gift to me and I'm thankful to have once again seen our world through their eyes. They restore my faith in the family's future. You and Ed have been so wonderful to share them with me so unselfishly."

THE FINAL CHAPTER

Jackie's Illness

In February 1994, it was announced that Jacqueline Kennedy Onassis had been diagnosed with non-Hodgkin's lymphoma. Treated with radiation and chemotherapy (and emergency surgery for a bleeding ulcer in April), her cancer outpaced the oncologists. Released from New York Hospital, Cornell Medical Center, she returned to her apartment, where she passed away on Thursday, May 19, 1994.

John F. Kennedy Jr. made a statement to the press the next morning:

"Last night, at around ten-fifteen, my mother passed on. She was surrounded by her friends and her family and her books and the people and the things that she loved.

"And she did it in her own way and in her own terms, and we all feel lucky for that and now she's in God's hands.

"There's been an enormous outpouring of good wishes from everyone both in New York and beyond. And I speak for all of our family when we say we're extremely grateful. Everyone's been very generous. And I hope now that, you know, we can just have these next couple of days in relative peace."

After a funeral mass at the Church of St. Ignatius Loyola, she was buried beside President Kennedy at Arlington.

★

After the diagnosis of her cancer, she spoke to a friend:

"Well, I have to make the best of the situation, and I'm going to do that."

★

She said to another friend a few months before her death, "But even if I have only five years, so what, I've had a great run."

*

She also said of her illness: "I don't get it. I did everything right to take care of myself and look what happened. Why in the world did I do all those push-ups?"

*

Just a month before she died, she told a friend that things were going well:

"I'm almost glad it happened because it's given me a second life. I laugh and enjoy things so much more."

Jackie's Will

In her last will and testament, Jacqueline Kennedy Onassis left:

"Copyright interest in personal papers, letters or other writings by me including any royalties" and "all tangible personal property including, without limitation, my collection of letters, papers and documents, my personal effects, my furniture, furnishings, rugs, pictures, books, silver, plates, linen, china, glassware, objects of art, wearing apparel, jewelry, automobiles and their accessories and all other household goods to my children."

*H*er final request of Caroline and John was to protect her privacy after her death:

"I request but do not direct my children to respect my wish for privacy, with respect to . . . papers, letters and writings . . . [and] take whatever action is warranted to prevent [their] display, publication or distribution."